Electrical installation work

SIXTH METRIC EDITION

strath

Electrical installation work

SIXTH EDITION

T. G. Francis

Revised by R. J. Cooksley
B Ed, MASEE, MIEEO.

Head of the Electrical Engineering Department and
School of Horology Hackney College

Longman
Scientific &
Technical

Longman Scientific & Technical
Longman Group UK Limited
Longman House, Burnt Mill, Harlow
Essex CM20 2JE, England
and Associated Companies throughout the world

First published 1950
Second edition 1953
Third edition 1958
Fourth edition 1968
Fifth metric edition 1971
Sixth edition 1988

British Library Cataloguing in Publication Data
Francis, T. G.
    Electrical installation work. – 6th metric ed.
    1. Electric wiring, Interior   2. Buildings
    – Electric equipment
    I. Title   II. Cooksley, R. J.
    621.319′24   TK3271

    ISBN 0–582–41344–3

Produced by Longman Singapore Publishers (Pte) Ltd.
Printed in Singapore

Contents

Preface to the sixth edition

The need for this sixth edition may be attributed to the following factors:

1. The publication in March 1981 of the Institute of Electrical Engineers *Wiring Regulations* (15th edition).
2. The re-structuring of the City and Guilds of London Institute's Electrical Installations courses, syllabi and examination methods.
3. The publication in December 1982 of the Joint Industry Board's (JIB) *Industrial Determination*. This scheme dictates the criteria for the education and training (the latter, both ON and OFF the job) for new entrants to the electrical contracting industry.
4. The continual development of new methods, materials, accessory and equipment design has brought about change in installation techniques and practice.

It is for those electrical installation students and others working within the installation industry that this book has been written. The objectives are to both complement the work conducted in further education colleges and other institutes, together with the updating of previous knowledge, techniques and methods.

The author is indebted to those institutes, organizations and companies who have so kindly granted permission for copyright materials and illustrations to be included or quoted in the text.

In particular:

1. The IEE for permission to quote from *Wiring Regulations* (15th edition). It should be stressed that the view expressed in this book related to the '15th edition', are those of the author and not those of any institute or organization that the author is or has been connected with.
2. The City and Guilds of London Institute for permission to reproduce recent and current materials and questions from their publications and examinations.

T. G. FRANCIS

List of symbols

The system of units used in this book is the International System of Units known as SI. The units for technology are based on the following six basic defined units:

Quantity	Symbol	Unit name	Unit symbol
Length	l	metre	m
Mass	m	kilogramme	kg
Time	t	second	s
Electric current	I	ampere	A
Thermodynamic temperature	T	kelvin	K
Luminous intensity	I	candela	cd

The following list of supplementary and derived units is based upon the British Standard 5775 1982, and upon Symbols and Abbreviations for use in Electrical and Electronic Engineering Courses (1968) published by the Institution of Electrical Engineers.

Quantity	Symbol	Unit name	Unit symbol
Area	A	square metre	m^2
Angle, plane	α, β	radian	rad
Angle, solid	ω	steradian	sr
Capacitance	C	farad	F
Current density	J	ampere per square metre	A/m^2
Density	ρ	kilogramme per cubic metre	kg/m^3
Energy	W	joule	J
		kilojoule	kJ
		watt hour	Wh
		kilowatt hour	kWh
Electric charge	Q	coulomb	C
Electric potential	V	volt	V
Electromotive force	E	volt	V
Force	F	newton	N
Frequency	f	hertz	Hz
Heat, quantity	Q	joule	J

Quantity	Symbol	Unit name	Unit symbol
Heat transfer coefficient	U	watt per square metre degree C	$W/m^2\,°C$
Illumination	E	lumen per square metre, lux	lm/m^2
		lux	lx
Impedance	Z	ohm	Ω
Inductance	L	henry	H
Luminous flux	Φ	lumen	lm
Magnetic field strength (magnetizing force)	H	ampere per metre	A/m
Magnetic flux	Φ	weber	Wb
		milliweber	mWb
Magnetic flux density	B	tesla	T
Magnetomotive force	F	ampere	A
Power, active	P	watt	W
		kilowatt	kW
Power, apparent	S	voltampere	VA
		kilovoltampere	kVA
Power, reactive	Q	voltampere reactive	VAr
Potential difference	V	volt	V
Pressure	p	newton per square metre	N/m^2
Quantity of electricity	Q	coulomb	C
Reactance	X	ohm	Ω
Reluctance	S	ampere per weber	A/wb
Resistivity	ρ	ohm metre	$\Omega\,m$
		microhm millimetre	$\mu\Omega\,mm$
Resistance	R	ohm	Ω
Specific heat capacity	c	joule per kilogramme degree Celsius	$J/kg\,°C$
Temperature, absolute	T	degree kelvin	K
interval		degree Celsius	°C, K
scale		degree Celsius	°C
Thermal conductivity	λ, k	watt per metre degree Celsius	$W/m\,°C$
Torque	T	newton metre	N m
Velocity, linear angular	v	metre per second	m/s
angular	ω	radian per second	rad/s

Decimal multiples and sub-multiples of units

Symbol	Prefix	Multiplying factor
T	tera	10^{12}
G	giga	10^{9}
M	mega	10^{6} (as in megavolts MV)
k	kilo	10^{3} (as in kilogrammes kg)
m	milli	10^{-3} (as in milliohms mΩ)
μ	micro	10^{-6} (as in microfarads μF)
n	nano	10^{-9}
p	pico	10^{-12}

Abbreviations for words

The following well-known abbreviations may be freely used:

Term	Abbreviation
Alternating current (as adjective only, e.g. a.c. supply)	a.c.
Direct current (as adjective only, e.g. d.c. circuit)	d.c.
Electromotive force	e.m.f.
Magnetomotive force	m.m.f.
Phase (as adjective only, e.g. 3-ph supply)	ph
Potential difference	p.d.
Power factor	p.f.
Revolutions per second	rev/s
Root-mean-square	r.m.s.

Introduction

An electrical installation in a building comprises various kinds of electrical apparatus fixed in position ready to use, together with the necessary connecting conductors, cables, fuse and control gear.

The local Electricity Supply Authority will normally install a supply cable to a suitable position inside the building, terminating the cable into a service board or 'cut-out'. This service is now termed the *origin of supply* and houses the fuses which are intended to protect the supply authority's equipment and metering facilities.

In order protect the building and its contents from the risk of fire and to safeguard the occupants or users of electricity from electric shock and burns, certain regulations have been formulated which control the design and erection of the electrical services. Certain of the differing sets of regulations are statutory and must be complied with. Typical statutory regulations are:

(a) Electricity Supply Regulations 1937
(b) Electricity (Factories Act) Special Regulations 1908 and 1944
(c) Coal and Other Mines (Electricity) Regulations 1956
(d) Building Standards (Scotland) Regulations with Amendment Regulations 1971–1980.

Readers should refer to Appendix 2 of the IEE *Wiring Regulations* (p. 108, 15th edn) for the more detailed list.

IEE Wiring Regulations

The *Regulations for Electrical Installations* (15th edn 1981), issued by the Institute of Electrical Engineers, lay down the requirements that have to be met by the designer of the installation and in turn by the electrical contractor/electrical operatives installing the electrical services, plant and/or equipment. The inspection and testing of an electrical installation is also conducted in accordance with these regulations.

Compliance with Chapter 13 of the IEE *Wiring Regulations* (15th edition) will generally satisfy the requirements of the previously mentioned statutory regulations. The reader should note also that the local fire officer, together with the electrical engineer for the company insuring a building may also impose their requirements which will be *in addition* to any of the statutory or IEE Regulations.

The method of complying with the IEE *Wiring Regulations* (15th edition) and their explanation is an inherent feature of this book, but it is the opinion of the author that every electrical installation operative and student should have those regulations together with a good working knowledge of them. Those operatives employed by organizations and firms coming under to the Joint Industries Board (JIB) were issued with a copy of the *Electrician's Guide to Good Electrical Practice* published by the Electrical, Electronic, Plumbing and Telecommunication Union (EEPTU).

Electricity generation, transmission, and distribution

A few notes on the above will serve to give the reader a picture of the 'background' to installation work.

Generation

Coal-fired generating stations.   The bulk of the electricity used in this country is generated in large coal-fired generating stations for reasons of economy in fuel, plant, and manpower. The station plant consists of furnaces, steam boilers, turbo-alternators, switchgear, transformers, and auxiliary equipment.

Coal is burned in the grates of large water-tube boilers which produce high-pressure steam. This steam is passed into steam turbines, which rotate at speeds of 50 or 25 revolutions per second. Each turbine is mechanically connected to an alternating current 3-phase generator, the combined machine being known as a turbo-alternator. The voltages of generation are 6600, 11 000, 22 000, and 33 000 volts, 3-phase.

It is clear from the foregoing that there are three transformations of energy. The energy in the coal is changed into heat energy in the steam, thence into mechanical energy of rotation and lastly into electrical energy.

The electrical energy from the alternators passes into the station busbars via switchgear, whence it is fed into outgoing cables or lines. Transformers are used to produce higher or lower voltage as required.

Nuclear power stations.   An increasing number of very large power stations have been, and are being constructed, in which the steam boilers are heated by means of nuclear reactors.

Oil-fired power stations.   In these stations coal is replaced by fuel oil as the heating agent for the boilers.

Hydro-electric power stations.   A number of these stations have been constructed in the North of Scotland, and Wales, whereby water power is used to drive water turbines which in turn drive the alternators.

Transmission.   When electrical power is carried in bulk by underground cables or overhead lines over considerable distances, it is said to be transmitted and the lines are called transmission lines. The outstanding example of transmission lines in this country is the 'British Grid'.

Alternating current distribution.   If power is to be supplied to various consumers locally, it is said to be distributed.

Distribution at high voltage involves the provision of transformer substations where the energy is transformed down to the standard low voltage range.

The definition of low voltage given in the IEE *Wiring Regulations* (p. 2) is:

'Normally exceeding extra-low-voltage (up to 50 V a.c. or 120 V d.c. either between conductors to earth) but not exceeding 1000 V a.c. or 1500 V d.c. between conductors or 600 V a.c. or 900 V d.c. and conductor and earth.'

To domestic consumers distribution is at low voltage. Electricity from the transformer substations is carried by feeder cables to suitable feeding points, whence cables called distributors radiate and run along the streets. The connection to any one house or shop is made by a service cable, which is tee-jointed to the distributor at a convenient position.

The various methods by which alternating current is distributed are:

Single-phase 2-wire
Single-phase 3-wire

Three-phase 3-wire
Three-phase 4-wire

The most widely used method is the 3-phase 4-wire.

These methods and earthing arrangements will be discussed in detail in Chapter 1.

The British Grid.   Prior to 1926, public electricity supply consisted of a large number of electrical supply authorities, each owning one or more generating stations and serving only a restricted area. There was but a limited amount of interconnection between areas. The Electricity (Supply) Act of 1926 established the Central Electricity Board with the responsibility of building and operating a national system of main transmission lines. These transmission lines and auxiliary equipment now in being are for the purpose of interconnecting the energy outputs of all the authorized electricity suppliers in the country.

Very many of the smaller power stations have closed down, and the remainder, together with a number of new large stations, have taken over the whole of the load.

Among the advantages of the interconnection are:

1. Large generating plants operate more efficiently than smaller ones.
2. Less reserve plant is needed.
3. In case of a plant breakdown in any one station, the local load may be supplied from the Grid.
4. Many more districts have been opened to electricity supply by tapping the secondary transmission lines.

The primary lines are 3-phase at 132 000 volts, 50 Hz (cycles per second), with secondary lines of 66 000 volts and 33 000 volts. A Super Grid at 275 000 volts has been constructed, and the greater part of this is being uprated to 400 000 volts. The bulk of the lines are overhead, but there are a few miles of underground cable.

Open-air substations, consisting of transformers and switchgear, connect the generating stations to the Grid.

The Electrical Council.   This Council advises the Government on questions affecting the supply industry, and also is required to promote and assist in the development of electricity supply.

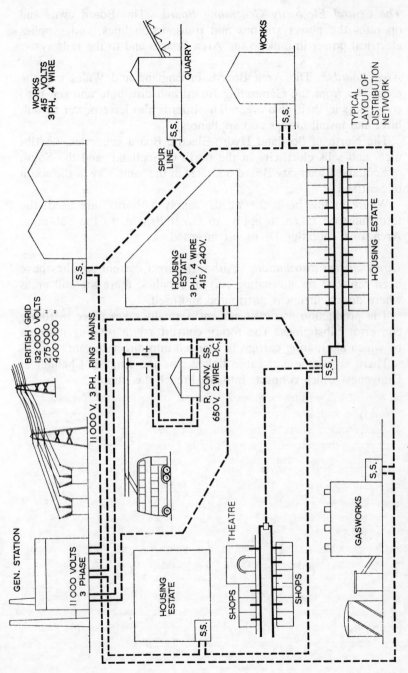

Fig. 1.1 Typical town distribution network

The Central Electricity Generating Board.   This Board owns and operates the power stations and transmission lines, and supplies electrical power in bulk to the Area Boards and to the railways.

Area Boards.   The Area Boards in England and Wales buy the electricity from the Generating Board and distribute and sell it to consumers in their own areas. The Boards also have power to sell, hire, and install fittings and appliances.

The North of Scotland Hydro-Electric Board generates, distributes, and sells electricity in the north of Scotland, and the South of Scotland Electricity Board operates in the same way in the south of Scotland.

Where in this book the words 'supply authority' are used, the term must be taken to apply, in Great Britain, to the Sub-area, Area, or Generating Division concerned.

Direct current distribution.   Although direct current supplies have been replaced by alternating current supplies, there are still areas where direct current is distributed and used.

The production of direct current normally takes place in rotary convertor substations. The rotary convertor is a rotating machine by which alternating current is changed into direct current.

There is also a d.c. cable link across the English Channel at Dungeness which connects into the French d.c. supply.

1 Supplies and systems

Standard voltages

When a supply authority's terminals to a consumer's premises are energized, an electric current will flow from one to the other around any closed circuit between them. This current is due to the difference of electric potential maintained between the terminals by the supply authority. This difference of potential is measured in volts, and is usually referred to as the SUPPLY VOLTAGE. The various ranges of voltage are defined in the Electricity Supply Regulations and the reader should note that there is some difference between this voltage range and that defined in the IEE *Wiring Regulations*.

Within the Electricity Supply Regulations, the voltage ranges are:

Low voltage	Not exceeding 250 volts
Medium voltage	Above 250 volts but not exceeding 650 volts
High voltage	Exceeding 650 volts

As an alternating current (a.c) supply is used as standard, the supply voltage will pass through a complete series of positive and negative values (cycles), a given number of times every second. This number is known as the FREQUENCY of the supply.

The international unit (SI) of frequency is the hertz (Hz), where 1 Hz equals 1 cycle per second.

The standard frequency used within the UK is 50 Hz and all equipment, plant and machines are designed to operate at this frequency. This is true for Europe and many other countries throughout the world, but in the USA the supply frequency is 60 Hz. It is important therefore that, before connecting any item of equipment to an electrical source, one should check that it is suitable for both the voltage (a.c. or d.c.) and the frequency of the supply. Whilst there are few direct current (d.c.) supplies outside industrial installations, and therefore most equipment is designed and made for a.c. supplies only, one must check as stated that the equipment is suitable, otherwise a dangerous condition may arise.

Variations of voltage and supply

It is impractical for a supply authority to maintain the same constant voltage at the origin of supply (service) of all its consumers due to the variation of voltage drop in the distribution cables as the connected loads vary.

Supply authorities are therefore permitted within the supply regulations to allow consumers' voltage to vary by not more than 6 per cent above or below the declared voltage. The permitted variation in frequency is 1 per cent above or below the declared frequency.

These declared voltage and frequency values are but two of the 'characteristics' that have to be obtained from the supply authority, when making an application for a supply. From Regulation 313-1 it will be seen that there are six items of information required:

1. Nominal voltage.
2. Nature of current and frequency.
3. Prospective short-circuit current at the origin of the installation.
4. Type and rating of the overcurrent protective device acting at the origin of the installation (service fuses).
5. Suitability for the requirements of the installation, including the maximum demand.
6. The phase-earth loop impedance of that part of the system external to the installation (supply network).

With items (3) and (6) these values are maximum values rather than the values that may be obtained by measurement.

Due to the definition within the IEE *Wiring Regulations* (15th edn) the word system has a somewhat different meaning than before when applied to electrical supplies.

An electrical system now comprises both the supply and the installation within a building. The type of system in use within a building is identified by the means of earthing that is employed, and again this type of information must be obtained from the supply authority when requesting a supply.

It will be seen from Part II (p. 12) and Appendix 3 (p. 110) of the IEE *Wiring Regulations* that some five systems are available, these being TN-C, TN-S, TN-C-S, TT and IT. This book will deal with only the three systems that are more commonly met, TT, TN-C-S and TN-S. The systems are designated by the initials used and these initials relate to the method of earthing provided by the supply authority, the relationship between exposed conductive parts and earth, and the arrangement of neutral and protective conductors.

Figures 1.2, 1.3, 1.4 show the earthing arrangements for the three systems mentioned, but the reader will need to understand why the initials used are grouped as they are.

TT system

This system is commonly met in rural districts where the incoming supply is by overhead cables supported on insulators mounted on poles. For the smaller installation only two conductors are used, the phase and neutral. This supply will have the neutral connected to earth at the supply transformer together with the metalwork of all casings or enclosures of the electrical equipment.

Therefore the first letter or initial T signifies that the supply has been earthed (T being used as it comes from 'terra firma' – meaning earth). The second letter T indicates that the installation has its own earth connection which is independent of the supply. This earth takes the form of an earthing conductor which is connected to an earth electrode (driven or buried in the ground) via an item of protective equipment termed a RESIDUAL CURRENT DEVICE. This RCD will be discussed in greater detail later, but in general terms is an automatic electro-mechanical switch which will switch off the supply when an earth fault current of a maximum value 30 mA or 0.030 A flows.

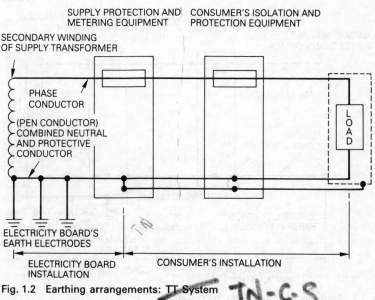

Fig. 1.2   Earthing arrangements: TT System   ~~TN-C-S~~

This diagram should be switched with 1.3.

TN-C-S system

This system uses the supply method which is now being installed in new installations. The earthing conductor is connected to a nickel-plated link which sticks out from the side of the service head. This link is bolted directly onto the neutral of the incoming supply cable, thus the supply neutral acts as the protective conductor as well, thereby cutting the costs of distributor cables for the supply authority.

Under earth fault conditions now, as the consumer's installation is earthed directly to the incoming neutral, all such faults become phase to neutral faults which allow a much larger fault current to flow thus operating protective devices (fuses, circuit breakers) much more effectively.

So again the first letter T indicates that the supply is earthed whilst the second and third letters N and C show that the supply neutral and protective conductor is combined in one. This conductor, therefore, is sometimes referred to as a PEN conductor (protective earthed neutral) as it combines the functions of both a protective conductor and neutral. The fourth letter S shows that neutrals and protective conductors are entirely separate from one another within the consumer's installation, all conductors being insulated.

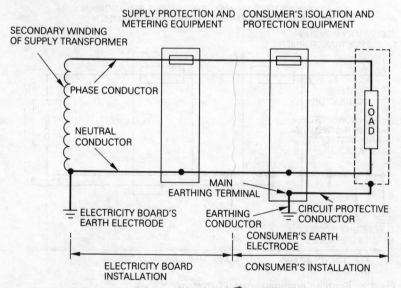

Fig. 1.3 Earthing arrangements: TN-C-S System

TN-S system

This is the system that has been commonly used since 1966 when the previous set of IEE Regulations were published (14th edn). At this time the supply authorities agreed to provide an earthing terminal to which the consumer may connect his/her earthing conductor.

This was normally effected by a connected to the lead sheath of the supply cable that was commonly used. However, as cable costs have escalated this particular cable type is rarely used for such distribution networks, giving way to a p.v.c.-insulated aluminium cable type.

The first letter T again indicates that the supply is earthed, whilst the second and third N-S show that the supply has a neutral which is separate from the protective conductor. The neutral being insulated from the lead sheath of the cable.

All such cable and earthing equipment and terminology will be explained in much greater detail in a further chapter in this book.

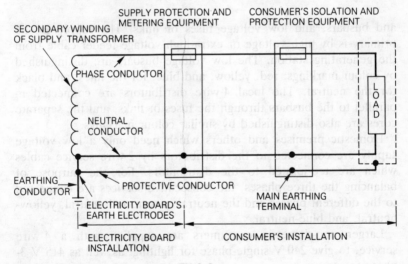

Fig. 1.4   Earthing arrangements: TN-S System

Three-phase 4-wire distribution

Figure 1.5 shows part of a typical 3-phase 4-wire distribution system. To supply the load in a given area, transformer substations are built at or near the load centre of the area. The substation contains high voltage switchgear, a step-down transformer, low voltage switchgear

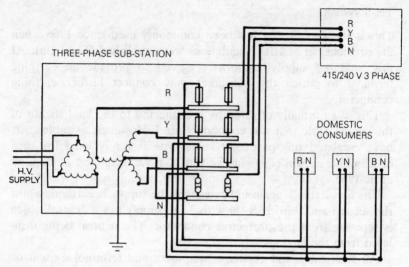

Fig. 1.5 Typical 3-phase 4-wire distribution

and busbars, and low voltage fuses or links. The supply to the primary is by high voltage or extra-high voltage feeder cable from the generating station. The low voltage busbars are distinguished by colour markings: red, yellow, and blue, for the lines, and black for the neutral. The local 4-wire distributors are connected in parallel to the busbars through the fuses or links, and the separate cores are also distinguished by similar colour markings.

Domestic premises and others which need only a low voltage supply are connected to the distributors by 2-wire service cables which are tee-jointed to the distributors. For the purpose of balancing the three phases, the consecutive services are connected to the different phases and the neutral in turn: red-neutral, yellow-neutral, and blue-neutral.

Larger non-domestic consumers are supplied with a 4-wire service, to give 240 V single-phase for lighting, as well as 415 V 3-phase for power.

2 Consumers' circuits

Internal distribution

It may be taken as a general statement that all types of load in a 2-wire installation, lights, heaters, motors, etc., are connected in parallel at the same voltage. Also, if a number of loads be supplied at the same voltage, the energy used by any one load is the same whether the others are connected or not. The addition or subtraction of other loads in parallel has no effect upon the particular load under consideration.

If the resistances of the connecting leads are taken into consideration, as they must be in detailed calculation, these two statements must be modified; thus, the arrangement of loads is a mixture of parallel and series-parallel circuits, and the potential differences across the separate loads vary in some small degree.

An internal distribution system consists in the connection in parallel of a group of loads in a final circuit, and the connection of this final circuit to local distributing busbars in a distribution board. Other groups of loads are connected to the same busbars, so that the groups are in parallel. In a larger installation a number of these busbars or fuseboards may themselves be connected in parallel to a main set of busbars, or a main distribution board, from which a pair of leads are connected to the supply service cable through suitable switchgear.

Each individual load may be separately controlled by a switch in its own circuit, as shown in Fig. 2.1. The separate groups are controlled by fuses or circuit-breakers. Fuses are shown in the diagrams (Figs. 2.1 to 2.6). Similar control is repeated right back to the main supply terminals. Thus, any individual load, or any group, or the whole is controlled by circuit-breaker, switch, or fuse. The above general system applies to all types of installation, large or small.

Figure 2.1 is a diagram of a simple 2-wire installation, supplying low voltage to a small lighting installation. The order of the control

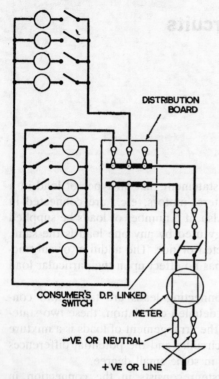

Fig. 2.1 Simple 2-wire installation

is shown. Service cable, sealing end-box, service fuse and neutral link, and meter are the property of the supply authority, and are sealed. The consumer's linked main switch, fuse and link, and distribution board follow in order. The final circuits are each connected to one way on the distribution board. The lights, each controlled by a 1-way switch, are connected in parallel.

Such '2-wire' installations are known as single-phase installations or circuits having one phase (red, yellow or blue) and a neutral. Readers should note from Table 52A (p. 68) 15th edition that whilst a single-phase circuit may be supplied from the red, yellow or blue phase of a supply, the final circuits from the distribution fuse board must be wired with red-coloured phase conductors and of course black for the neutral. In order to identify switchlines at any switch position where a twin or multicore sheathed cable has been installed, the switchline should be provided with a red sleeving.

Figure 2.2 shows how the various final circuits supplying different

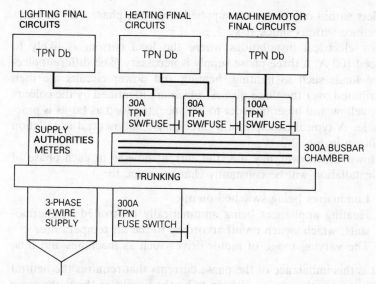

Fig. 2.2 Typical 3-phase 4-wire installation

loads would be distributed over the three phases in order to 'balance' the load as is practicable. In practice, due to the changing current demand brought about by luminaires being switch on/off, thermostats controlling heating loads and the varying usage of lift motors, etc. there is always some imbalance between the phases. It is this imbalance that requires the neutrals in a consumer's installation to be the same size cross-sectional area as that of the phase conductors. Readers should note that both phase and neutral conductors are defined as LIVE conductors in the 15th edition.

In an installation that is supplied from more than one phase, the risk of serious or fatal shock conditions are increased. It is for this reason then that the IEE Regulations require certain precautions to be taken (Reg. 514-4). The voltage between any two phases will be in the order of 415 V (above 250 V) so that if an item of equipment, switchgear or an enclosure within which a voltage above 250 V exists, a warning notice must be reliably fixed on or adjacent to the equipment, etc. in order that a warning is given before opening the enclosure and gaining access to live parts. This warning notice must indicate the maximum voltage present and is also required when separate items of equipment, supplied from different phases, can be simultaneously reached by a person.

Generally it is good installation practice to ensure that all socket outlets within one room are supplied from one phase only in order to reduce serious shock risk.

For electrical installations where the load current is likely to exceed 100 A, a three-phase supply is necessary. The different electrical loads such as lighting, heating and power circuits are then distributed over the three phases, which are identified by the colours red, yellow and blue, in order to 'balance' the load as far as is practicable. A typical example of a three-phase and neutral installation is shown in Fig. 2.2.

However, in everyday use, the current flowing in each phase of an installation will be constantly changing, due to:

(a) Luminaires being switched on/off;
(b) Heating appliances being automatically controlled by thermostats, which switch on/off according to the air temperature;
(c) The varying usage of motor drives such as machines and lifts.

It is this imbalance of the phase currents that requires the neutral conductors within an installation to be the same size (have the same cross-sectional area in mm^2) as that for the phase conductors.

Readers should note that both phase *and* neutral conductors are defined as *live* conductors, in the IEE Wiring Regulations (15th edn).

In a three-phase four-wire installation, as the voltage between each phase will be in the order of 415 V, the risk of serious or fatal electric shock conditions are increased. It is for this reason that the IEE wiring Regulation 514-4 requires certain precautions to be taken.

If an item of equipment, switchgear or an enclosure is supplied from more than one phase, the voltage available will exceed 250 volts. A warning notice which indicates the maximum voltage present, must be reliably fixed adjacent to or on the equipment etc., in order that a warning is given before opening the enclosure and gaining access to live parts. The same warning notice is required when separate items of equipment, supplied from different phases, can be simultaneously reached by a person.

In order to avoid this same shock risk, it is good installation practice to ensure that all socket outlets within one room are supplied from one phase of the supply only.

This difference in phase-to-phase and phase-to-neutral voltage within a three-phase four-wire installation may be seen in Fig. 2.3

Figure 2.3 also shows the 'star' connected secondary windings of a supply transformer, which would normally be installed within the

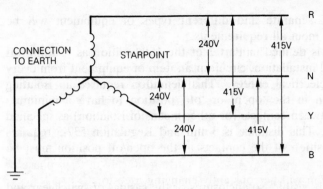

Fig. 2.3 Voltages obtainable from 3-phase 4-wire distribution

supply authority's 'substation', that the common or star point connection is connected to the mass of earth. This is normally achieved by connection to rods, tapes or plates buried directly in the ground adjacent to the substation.

From this star point an earthed conductor, called the neutral, is brought out with the red-, yellow- and blue-phase conductors. This method of connection has the advantage of providing two alternative voltages within an installation; 415 V between any two phases (R-B, B-Y, B-N) and 240 V between and phase and neutral conductor (R-N, Y-N, B-N).

Lighting and other small loads such as immersion heaters and cooker circuits use 240 V, while the 415 V 3-wire circuit would supply motors driving rotating plant in factories and for lifts in office blocks, etc.

The 415 V 4-wire (TPN) method is widely used for the distribution of supplies within commercial and small industrial installations.

Control, distribution and overcurrent protection of a consumer's installation

Every installation supplied from an external source has to be controlled by switchgear incorporating the following:

1. A means of isolating the whole of the installation.
2. Overcurrent protection (overload and short-circuit).
3. Protection against earth leakage current.

With a simple installation items 1 to 3 may be incorporated in a single item of switchgear, but with the larger or more complex installation, separate and different types of equipment will be required to meet all requirements.

Isolation is defined in Part 2 of the 15th edition as 'Cutting off an electrical installation, circuit or an item of equipment from every source of electrical energy.' This definition requires an isolating device when in the 'open' or 'off' position to have a minimum distance between contacts (or other means of isolation) as specified in BS 5419. This distance is 3 mm and Regulation 537-5 requires that the position of the contacts in the open/off position must be either:

1. Externally visible, so enclosures or the casings of switchgear and control gear must be transparent or have a small window provided, or
2. Have a reliable means of indicating when the contacts have reached the required separable distance. This will entail having the indicating device mechanically attached to the moving contacts of the isolating device.

Access to isolators   Within the *IEE Wiring Regulations* (15th edn) Chapters 46, 476 and 537 detail other requirements for isolation, and are such that access to such isolators is restricted to only authorized persons. For the larger installation this will mean a locked switch-room with access for only skilled or instructed persons (as defined in Part 2 of the 15th edition). If this is not the case then such isolators must incorporate a means of locking the switches in the OFF position. This will normally require a padlock or having a removable handle to the switch. Regulation 476-4 requires that each such padlock or handle shall not be interchangeable with any other isolation.

Overcurrent protection

The domestic installation generally has a 'Consumer's Control Unit' (CCU) which contains the main isolating switch and final circuit fuses or miniature circuit breakers. There is some doubt as to whether the double-pole switch used as the isolator now meets the requirements of the 15th edition regarding:

1. Separable distances of switch contacts in the off position.
2. Reliable indication of this separable distance.
3. The risk of the switch being accidentally reclosed by mechanical shock or vibration.
4. The short-circuit current rating of the switch.

The live conductors forming the circuits which go to make up the whole of an installation must be protected by one or more devices which will automatically disconnect the supply in the event of an overcurrent flowing.

Readers should note that an overcurrent may be due to either an overload or short-circuit condition, and it is important to realize the difference between the two.

Overload protection

With an overload the circuit, although electrically sound and without having a fault, may be subjected to an overcurrent due to having too many electrical appliances or loads switched on at the same time. The current that will flow under this condition will exceed that for which the conductors and circuit fuse or circuit breaker was selected. This overload current must be disconnected before it could cause a temperature rise which may damage insulation, joints, terminations or any materials close to the conductors.

Co-ordination of protective devices

Regulation 433-2 requires the characteristics of the device (fuse or circuit breaker) protecting the circuit to be co-ordinated with the conductors forming the circuit such that:

1. Its norminal current (fuse rating) or current setting for a circuit breaker (I_N) is not less than the design current, which is the maximum normal circuit current (I_B).
2. I_N is not greater than the current-carrying capacity (I_Z) of the circuit conductors.
3. The current which operates the fuse or circuit breaker, causing it to disconnect the circuit (I_2), does not exceed 1.45 times the lowest of any of the current-carrying capacities (I_Z) for the circuit conductors.

In the note to this Regulation, it will be seen that the foregoing may be represented in formula

$$I_B \leqslant I_N \leqslant I_Z \text{ and } I_2 \leqslant 1.45 \, I_Z$$

Short circuits   A short-circuit fault is one where a fault results in neglible impedance existing between phase conductors or a phase conductor and neutral. In practice this means that two such conductors or live parts come into contact due to the absence or failure of insulation.

The resulting inrush of fault-current will be of an extremely high value and it is most important that this short-circuit current is disconnected in a very short time. When protection devices are dealt with in Chapter 5 it will be seen in greater detail that under short-circuit conditions there will be a high level of energy available to create heat and mechanical effects. This energy is known as 'let-through energy' and will be proportional to I^2t, where I is the short-circuit fault-current and t is the duration of the short circuit in seconds.

In practice then switchgear and protective devices must be selected after having determined that they are of the correct type and rating to withstand the effects of short circuit. Should this not be done, then extreme damage due to heating and electro-mechanical forces may occur.

Earth-fault current

It is also necessary to provide protection against earth leakage current. This is a current that can flow to the mass of earth via the exposed conductive parts of an installation. These might be metallic enclosures, conduits, cable trunking or sheaths of cables. Such current may also flow via extraneous conductive parts, which are formed by the metallic surfaces of other services.

If all such metalwork is not effectively connected to earth, then it is possible for that metalwork to be raised to some potential above earth; in practical terms it will be 'alive' and the risk of electric shock may exist.

Contact with such exposed and extraneous conductive parts is termed 'indirect contact' as the risk of shock only exists when a fault exists. The means of protection against indirect contact is termed **E**arthed **E**quipotential **B**onding and **A**utomatic **D**isconnection – or EEBAD for short.

Figure 2.4 shows how, in a domestic installation, exposed conduction parts can be bonded to:

1. Main water services.
2. Main gas services.
3. All other piped services and ducting.
4. The risers and ducting of central heating and air conditioning systems.
5. Any exposed structural steel work.

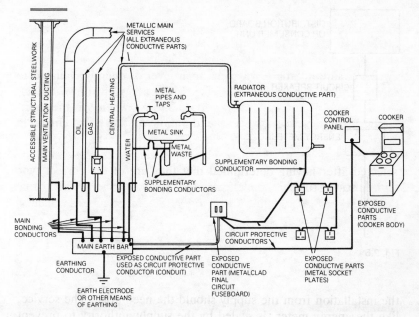

Fig. 2.4

It will be noticed that the main bonding conductors connect items 1 to 5 with the main earthing terminal. From Regulation 413-2, Note 1, it will be seen that this bonding is intended to create a ZONE within which any voltage between exposed and extraneous conductive parts is minimized.

In practical terms, when a phase conductor comes into contact with earthed metalwork, due to damaged or failed insulation, then the inrush of current will be of such a high value that the protective device for that circuit will operate within a specified maximum time and disconnect that faulty circuit. This is known as the phase earth fault.

The sequence of control equipment for a simple installation, possibly a domestic installation, is shown in Fig. 2.5. Readers should note that the service fuse within the supply authority's cut-out or service has a short-circuit rating of some 16 000 amperes (16 kA) and is intended to protect the supply authority's cable and equipment from any short-circuit fault that may occur in the consumer's installation. This service fuse is also used to disconnect

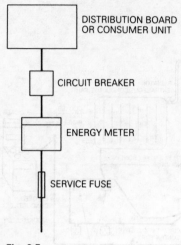

DISTRIBUTION BOARD
OR CONSUMER UNIT

CIRCUIT BREAKER

ENERGY METER

SERVICE FUSE

Fig. 2.5

the installation from the supply should the need arise. The service, like the energy meter, is sealed by the supply authority to prevent unauthorized tampering with the equipment and possible interference with the meter to avoid payment for the electrical energy.

Use of residual currents

It has become common practice to replace the 100/80 A double pole (DP) switch controlling the domestic consumer's control unit with a residual current circuit breaker. This RCD is designed to detect earth fault current and will trip when a fault current of 30 mA flows.

However, it should be pointed out that this practice impinges upon Regulation 314-1 which requires every installation to be divided into circuits so as to:

1. Avoid danger and minimize inconvenience in the event of a fault.
2. Facilitate safe operation, inspection, testing and maintenance.

With such an arrangement, shown in Fig. 2.6, it should be noted that an earth fault on the cooker or immersion heater circuit would cause the RCD to trip, thus isolating the whole of the installation from the supply. This could create danger in as much as there might be a risk of someone tripping or falling due to the loss of the lighting, should there be the need for artificial lighting.

Consumer unit practice

It is for this reason that some manufacturers are now marketing a consumer unit that has two banks of fuses or miniature circuit breakers (MCB). One bank protects lighting circuits and may be controlled by a 100/80 A DP switch, while the other bank protecting socket outlet circuits is controlled by an RCD.

It is good installation practice to select such a consumer's unit so that one or more spare circuit ways are provided for future extensions to the installation. But often in a competitive situation it is not done, as obviously an 8-way consumer's control unit (CCU) will cost more than a 6-way. It is this economic factor that cause some specifiers to continue using rewireable fuses and not MCB's; although the latter are much more convenient to reclose after a fault, they are considerably more expensive. For example, an 8-way CCU with RCD control of 6 MCB's and having two spare ways would cost in the order of four times that of a 6-way CCU with 80 A DP switch control of six rewireable fuseways.

Regulation 314-3 requires that the number of final circuits and the number of outlets supplied from any such final circuits have to be such that all the requirements for overcurrent protection, isolation

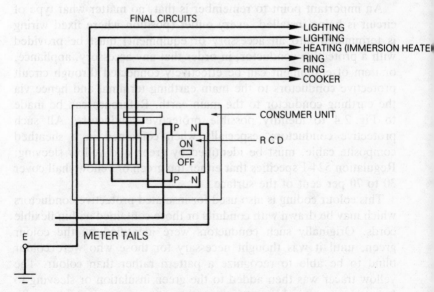

Fig. 2.6 Typical arrangement at a domestic consumer's unit

and switching and the current-carrying capacities of the circuit conductors are met.

Every installation must be comprised of more than one final circuit. As a minimum even the simplest installation should have two lighting circuits, so that with the loss of one due to a fault condition some of the lights will remain operational and thus avoid danger and inconvenience. Regulation 314–4 requires that each final circuit has its own phase and neutral conductors which are identifiable in the CCU or distribution fuseboard as terminating the neutrals in the same order as the respective phase conductors.

Figure 2.6 shows a small installation which is comprised of final circuits serving: Lighting; Water heating; Cooking appliances; Socket outlets. Others may include space or underfloor heating and specially protected circuits supplying lighting or socket outlets external to the building.

For the commercial or industrial installation, the variety of circuit designations may well be computer supplies, lifts, machine tools, welding sets, induction furnaces and many others.

Protective conductors

Within this book only the basic circuits are considered.

An important point to remember is that, no matter what type of circuit is being installed, every outlet (position where fixed wiring is terminated into an accessory or equipment) must be provided with a protective conductor, in order that any accessory, appliance, or item of equipment can be effectively connected through circuit protective conductors to the main earthing terminal and hence via the earthing conductor to the main earth. Reference can be made to Fig. 2.4 to identify possible protective conductors. All such protective conductors, especially when contained in a sheathed composite cable, must be identified by green and yellow sleeving; Regulation 524-1 specifies that either the green or yellow shall cover 30 to 70 per cent of the surface.

This colour coding is also used for insulated protective conductors which may be drawn with conduits or those contained within flexible cords. Originally such conductors were identified by the colour green, until it was thought necessary for those who were colour blind to be able to recognize a pattern rather than colour. The yellow tracer was then added to the green insulation or sleeving to form this recognizable pattern.

Wiring circuits for lighting

The method for wiring final lighting circuits may be either the loop-in, three-plate or the joint-box methods.

The loop-in method enables all joints or terminations to be made at ceiling roses, luminaires, switches or other accessories. Hence all such joints remain accessible for the purpose of alterations, additions or for testing.

The loop-in method is used with conduit or trunking installations and although more cable is used the avoidance of jointed conductors in boxes is seen as a big advantage.

The 3-plate method avoids the greater use of cable, as joints are made in the ceiling roses which have shrouded terminals (to BS 67). The phase conductors are joined here rather than 'looping in' at the switch positions. The 3-plate method is widely used on domestic lighting installations employing wiring systems such as p.v.c. sheathed p.v.c. insulated twin or 3-core with c.p.c.

Figure 2.7(a) shows the basic circuit diagram of a final circuit of seven lighting outlets – two controlled separately by 1-way switches, three controlled as a group by a 1-way switch, and two also controlled by a 1-way switch.

For the purpose of clarity, the c.p.c. and terminals have been omitted. Readers should also note that the flexible conductors used for connecting lampholders to the ceiling roses have to be of the circular sheathed type, the twin twisted type of flexible cord not being permitted.

Figure 2.7(b) shows the same circuit with the connections arranged for the 'looping' method. It should be noted that Regu-

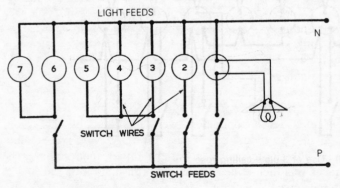

Fig. 2.7(a)   Lighting final circuit

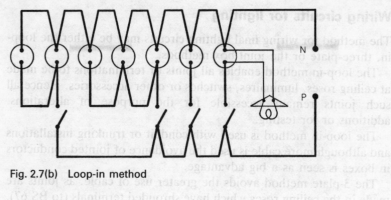

Fig. 2.7(b)   Loop-in method

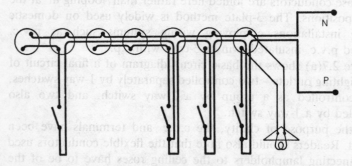

Fig. 2.7(c)

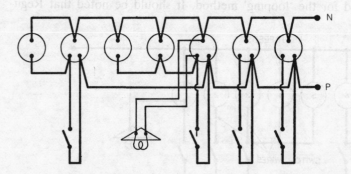

Fig. 2.7(d)   Use of 3-plate ceiling rose

lation 521-8 requires the method of wiring to be such that a phase conductor is NOT enclosed or surrounded by ferrous material (iron or steel). Otherwise the magnetic effect will be such that heating will occur.

Figure 2.7(c) shows again the same circuit, in suitable wiring form as it may well be wired in a conduit system with the 'switch drops' contained in a conduit taken from the circular conduit box (loop-in box).

Figure 2.7(d) shows the same circuit with the connections arranged for the 3-plate method. Figure 2.8 gives details of the connections at one ceiling rose.

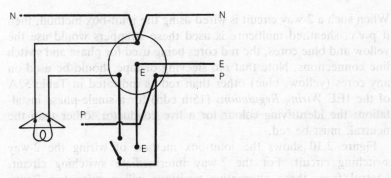

Fig. 2.8   One light wired from 3-plate ceiling rose

Two-way intermediate switching

When it is necessary to control a lighting outlet from two positions then a 2-way switching circuit may be utilized. When three or more control positions are necessary then a 2-way intermediate circuit will be required.

Figure 2.9 is the wiring diagram showing one lighting outlet controlled by two 2-way switches, such as may be required, say, in any room having more than one entrance door, in a hall or corridor, as well as on a staircase. Note that the 2-way switch has three terminals, one being the common, to which either the phase conductor A or the switch live A is connected. The other live terminals on each switch B,C and B,C are inter-connected by two conductors which give alternative connections depending on the switching mode of each switch. These two conductors are known as the 'strappers'.

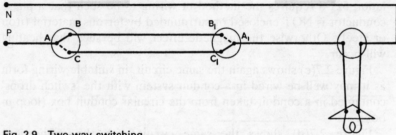

Fig. 2.9 Two-way switching

Use of joint boxes

When such a 2-way circuit is wired using the joint-box method, then if p.v.c. sheathed multicore is used these strappers would use the yellow and blue cores, the red cores being used for phase and switch line connections. Note that red sleeving or tape should be used on any cores (yellow, blue) other than red as indicated in Table 52A of the IEE *Wiring Regulations* (15th edn) for a single-phase installation; the identifying colour for a live conductor, other than the neutral, must be red.

Figure 2.10 shows the joint-box method of wiring the 2-way switching circuit. For the 2-way intermediate switching circuit, control from three alternative positions will require two 2-way switched and one intermediate switch as shown in Fig.2.11.

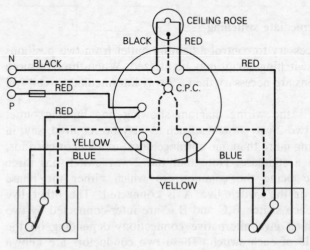

Fig. 2.10 Two-way switching

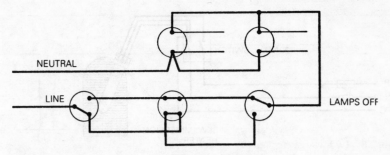

Fig. 2.11 Use of 2-way and intermediate switches

When additional control positions are required, then further intermediate switches must be connected into the circuit as shown in Fig. 2.12.

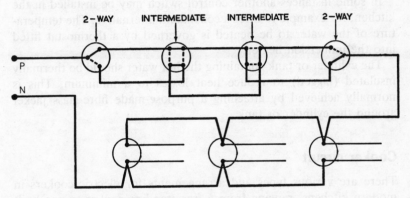

Fig. 2.12 2-way intermediate switching from four positions

Water heating circuits

In a domestic electric water-heating installation it is normal to install a 3 kW immersion heater, which is screwed into a boss fitted directly into the hot-water cylinder or tank. Such cylinders or tanks will have a capacity exceeding 15 litres. The immersion heater circuit must be separate from any other circuit. Such a circuit will be protected by a 15 A fuse or MCB and controlled by a 20 A double-pole switch having a neon indicator. Final connection to the heater from the DP switch is normally by 85 °C rubber insulated HOFR (heat, oil and flame retardant) sheathed 3-core flexible cord.

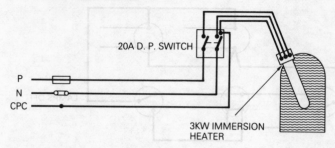

Fig. 2.13   Immersion heater final circuit

Note that an equipotential bonding conductor should be connected from the earthing terminal on the heater to the various pipe runs associated with the cold- and hot-water services. Fig. 2.13 shows such an immersion heater circuit.

In some instances another control switch may be installed in the kitchen, for example to offer a convenient alternative. The temperature of the water to be heated is governed by a thermostat fitted into the immersion heater.

The cylinder or tank containing the hot water should be thermally insulated (lagged) to reduce heat losses to a minimum. This is normally achieved by attaching a purpose-made fibre-glass jacket around the cylinder or tank.

Cooker circuit

There are various types and arrangements for electric cookers in modern kitchens, ranging from a free-standing cooker to one built into the kitchen units. Latterly it has become fashionable to have 'split level' cookers whereby a hob unit is built into the worktop, and the oven/grill unit is housed in the kitchen units.

Application of diversity factors

Whilst the average loading for a domestic cooker is some 13 kW, it is not unusual for the electrical loadings to exceed this. The current rating of the cooker circuit to be installed may be determined from an assessment of the current demand of the cooking appliance(s). This will entail allowances for 'diversity'.

Table 4A in Appendix 4 of the IEE *Wiring Regulations* (15th edn) indicates the method to be employed in order to assess the circuit rating using the appropriate diversity factors. These factors are based on the assumption that it would be unlikely for all the heating elements (rings, hotplates etc.) to be switched on at the same time. Even if this were to be so, it would again be unlikely for all simmerstat controls to be in the ON mode simultaneously.

For a domestic cooking appliance the current assessment is:

The first 10 A of the rated current plus 30 per cent of the remainder of the rated current plus 5 A if a socket outlet is incorporated in the cooker control unit.

EXAMPLE
Determine the circuit rating for a 240 V – 12 kW cooker controlled by a unit having a 13 A socket outlet incorporated.

$$\text{Rated maximum demand} = \frac{12\ 000\ \text{watts}}{240\ \text{volts}}$$

$$= 50\ \text{A}$$

$$\text{if Current (A)} = \frac{\text{Power (W)}}{\text{Volts (V)}} = \frac{12\ 000\ \text{watts}}{240\ \text{volts}}$$

$$\frac{12\ 000\ \text{watts}}{240\ \text{volts}}$$

Applying diversity

First 10 A	= 10 A
30% of 50–10	
$= \dfrac{30}{100} \times 40$	= 12 A
Plus 5 A for socket outlet	= 5 A
Total	= 27 A

Therefore a 30 A circuit could be installed. This is the norm for most domestic installations, although a 45 A circuit can be accommodated in all consumer units.

Every fixed or stationary domestic cooking appliance must be controlled by a double-pole switch separate from the appliance, and has to be within 2 m of the appliance. Such a switch may control both a hob unit and the oven unit providing that both appliances are not more than 2 m from the switch.

It is becoming common practice to use a 45 A DP switch incorporating a neon indicator lamp to control cooking appliances, rather than use cooker control units having a built-in 13 A socket outlet. Such a 45 A switch is physically smaller and neater in appearance, and with the increased provision of socket outlets on ring or radial circuits in kitchens today, there is no real need to have a socket outlet in the cooker control.

Readers should note that it is normal to ensure installation of equipotential bonding conductors between all exposed conductive parts (cooker frame) and extraneous conductive parts (sink top, piped services) that may exist in a kitchen. This is particularly necessary where electrical appliances such as washing machines, refrigerators, dish-washers, cookers together with portable appliances (food-mixers, toasters, irons, etc.) may be installed or used close to metallic sink tops. All such metallic sink tops, if conforming to British Standards, will have provision for terminating the equipotential bonding conductor. This conductor is also termed a 'supplementary bonding conductor'.

Final circuits for socket outlets

Before the late 1940s, it was common practice to install a maximum of one 15 A switched socket outlet per room, normally adjacent to the fireplace.

This necessitated the installation of a number of separate such circuits, supplied from a 15 A double-pole fuseboard and controlled by a 60/100 A switchfuse. Such equipment was both bulky and expensive.

In addition, 5 A or 2 A socket outlets were provided to supply appliances other than electric fires or radiators.

Towards the end of the 1940s the ring final circuit was introduced, although at that time the 13 A square-pin plug and socket to BS 1363, which is standard these days, was not used. The introduction of the ring final, NOT ring main which is unfortunately the wrong term used by many operatives and trainees, effectively provided a means of installing a greater number of socket outlets and in doing so:

1. Avoided the need for long trailing flexible cords and multiple adaptors, which one socket outlet per room seemed to encourage.
2. Reduce the material costs of providing more socket outlets per room.
3. Provide a more convenient siting of outlets and increased safety in use.

Ring final circuits

For a ring final circuit each circuit conductor shall commence from a fuseway (30 A) at the distribution board, looping into each socket outlet or fused spur unit and return to the same fuseway to form a closed ring. This allows for the load current at a particular socket outlet to be shared between both conductors supplying that socket outlet. If a separate circuit protective conductor is run, either within a sheathed cable or as a single insulated (green/yellow) conductor, then this conductor must also commence at the earthing terminal wtihin the distribution board, looping into all socket outlets and fused spur units before returning to the same earthing terminal, in exactly the same way as the wire conductors.

The socket outlets used for domestic ring and radial circuits will comply with BS 1363 and are shuttered to prevent accidental contact with live parts.

Table 5A and Appendix 5 in the *IEE Wiring Regulations* (15th edn) deals in some detail with all requirements for ring and radial circuits. Referring also to Fig. 2.14 it is essential that the following important items are remembered.

1. There may be an unlimited number of socket outlets connected to a single 30 A ring final circuit, provided that the floor area served by the ring does not exceed 100 m^2.

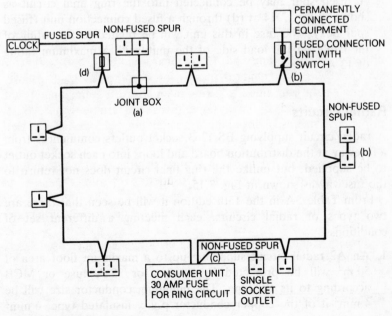

Fig. 2.14 Ring final circuit

2. Where the floor area exceeds 100 m², then additional ring final circuits should be installed, distributing the socket outlets over the ring final circuits so as to share the assessed load.

3. It is good practice to install a separate ring final circuit in the kitchen of a dwelling, as this is normally the 'load centre' of most domestic installations.

4. Branch circuits or unfused 'spurs' may be taken directly from the ring as shown in Fig. 2.14:

(a) by connection through a 30 A joint box;

(b) directly from the terminals of a socket outlet or fused connection box connected into the ring;

(c) directly from the same terminals within the consumer unit as the phase, neutral and c.p.c. of the ring connected.

With (a), (b) and (c) it is important to note that the same size of cable should be used as that of the ring conductors.

5. Such an unfused spur may supply one twin socket outlet or one single socket outlet or one item of permanently connected equipment.

6. The number of such non-fused spurs must not exceed the total number of socket outlets and any stationary equipment directly connected to the ring final circuit.

7. A fused spur may be connected into the ring final circuit as shown in Fig. 2.14 at (d) through a fused connection unit (fused spur box). The fuse in this unit will be related to the rating of the cable on the load side of the unit, with a maximum rating of 13 A.

Radial circuits

A radial circuit supplying BS 1363 socket outlets commences from a fuseway at the distribution board and loops into each socket outlet to be supplied, but unlike the ring final circuit does not return to the fuseway as shown in Fig. 2.15.

From Table 5A in the 15th edition it will be seen that there are two types of radial circuits, each meeting a different set of conditions.

1. An A2 radial circuit supplying up to a maximum floor area of 50 m² will be protected by a 30 A or 32 A fuse or MCB according to its BS type. The minimum conductor size will be 4 mm² if of the copper conductor p.v.c. insulated type, 6 mm²

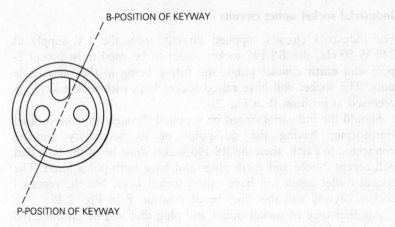

B-POSITION OF KEYWAY

P-POSITION OF KEYWAY

Fig. 2.15 BS 196 socket outlet

if copperclad aluminium with p.v.c. insulator and 2.5 mm^2 if mineral insulated copper conductors.

2. An A3 radial circuit supplying up to a maximum floor area of 20 m^2 will be protected by a 20 A fuse or MCB. The minimum conductor size will be 2.5 mm^2 if the copper conductor p.v.c. insulated type, 4 mm^2 if copperclad aluminium with p.v.c. insulator and 1.5 mm^2 if mineral insulated copper conductors.

Socket outlets

The socket outlets, either switched or unswitched which conform to BS 1363, are recognized by the three rectangular pins on the plugs which fit into the sockets. The plugs and sockets used for industrial installations have to conform to BS 196. Radial or ring final circuits may be employed and when assessing the current demand for such circuits, the maximum load current (design current) must not exceed 32 A which is the largest rating of fuse or MCB permitted.

The number of such socket outlets permitted is unlimited and any 'spurs' taken from a ring final circuit must be connected through a fused connection unit (spur box) or be protected by an MCB of not more than 16 A. Non-fused spurs are not used with this type of socket outlet.

Industrial socket outlet circuits

For industrial circuits supplied directly from the LV supply at 240 V–50 Hz, the BS 196 socket outlet to be used must accept 2-pole and earth contact plugs, the fusing being in the phase pole only. The socket will have raised socket keys with socket keyways recessed at position B in Fig. 2.15.

Should the industrial circuit be supplied through a double-wound transformer having the mid-point of its secondary winding connected to earth, then the BS 196 socket must be of the type that will accept 2-pole and earth plugs and have both poles fused. The socket outlet again will have raised socket keys, but the recessed socket keyway will this time be at position P in Fig. 2.15.

Another type of socket outlet and plug that will be encountered in industrial and certainly construction site installations conforms to BS 4343. These socket outlets and plugs are available in 16 A, 32 A, 63 A and 125 A ratings. The 32 A, 63 A and 125 A socket outlets may only be connected into a radial circuit supplying one socket outlet only. The 16 A rating outlet, however, may be connected into a 20 A radial circuit supplying an unlimited number of such outlets, consideration being taken of the assessed load that may be imposed on the circuit.

These socket outlets may be used on single- or three-phase supplies at voltages up to 750 V. Discrimination between the

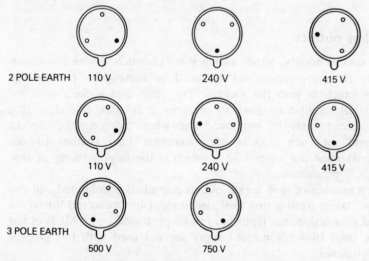

2 POLE EARTH   110 V   240 V   415 V

110 V   240 V   415 V

3 POLE EARTH   500 V   750 V

Fig. 2.16   BS 4343 plugs and sockets

various voltages that the outlets are operating on is achieved in
either of two ways:

1. By using a colour code.
2. By the location of the earth contact relative to a keyway.

The colour code for BS 4343 sockets and plugs is:

Violet	25 V
White	50 V
Yellow	110 V to 130 V
Blue	220 V to 240 V
Red	380 V to 415 V
Black	500 V to 750 V

Fig. 2.16 indicates the position of the earth pin/contact relative to
the keyway of BS 4343 sockets and plugs.

3 Conductors and cables

Cables form the necessary connections between the machine which generates electricity and the apparatus which uses it. They comprise a very wide range of sizes and types.

The necessary requirements of a cable are that it should conduct electricity efficiently, cheaply, and safely. To this end it should not be too small so as to have a large internal voltage drop. It should not be too big so as to cost too much originally, and the necessary joint boxes, etc., should not be too costly. Its insulation should be such as to prevent leakage of current in unwanted directions, and thus to minimize risk of fire and shock.

A cable has three main parts – the conductor, the insulation, and the mechanical protection.

Conductor materials

Copper and aluminium are the materials used as conductors in power and lighting cables. Copper has lower resistivity and thus higher conductivity than aluminium. This means that copper conductors have smaller cross-sectional area and take up less space than aluminium for the same current capacity. On the other hand, aluminium has about one-third the weight of copper and so will have an advantage in some circumstances.

Copper conductors may be annealed or hard-drawn. Annealed copper conductors are comparatively soft and pliable and are most suitable for indoor and outdoor wires and cables laid or fixed in position. Hard-drawn copper conductors, which have a very high tensile strength, are used as overhead wires mainly in the bare form. The great majority of cables in use are of copper.

With some insulation materials, but not all, copper conductors need to be tinned.

Aluminium conductors are made in all standard sizes but are used at present only in the larger sizes. IEE Reg 521-1 lays down that all cable conductors of cross-sectional area 10 mm^2 or smaller shall be of copper or copperclad aluminium.

The respective resistivities of copper and aluminium are

17.24 $\mu\Omega$ mm and 28.2 $\mu\Omega$ mm.

The measured resistances of made-up cable conductors are normally slightly higher than those calculated from the above resistivity values. For calculation purposes in this book therefore, values of resistivity for annealed copper and aluminium will be taken as 17.5 $\mu\Omega$ mm and 28.5 $\mu\Omega$ mm respectively.

Stranding

To ensure flexibility and ease of handling, conductors are stranded: a number of small wires twisted together spirally forming a core equivalent to a single wire of the required size. The numbers of strands used are 1, 7, 19, 37, 61 and 127. The sizes of conductors range from 1.0 mm^2 (1/1.13 mm) to 630 mm^2 (127/2.52 mm). The latter conductor, for example, consists of 127 strands of circular conductor, each strand of 2.52 mm diameter, with a total cross-sectional area of 630 mm^2. The complete range may be found in various tables in the Regulations.

Insulation

The function of the insulation is to confine the electricity to the conductor. To this end the insulation itself must have a very high resistance. For normal work the insulation is arranged to surround the conductor throughout its length. For overhead wires it is normally sufficient to provide insulation (e.g. a porcelain insulator) at the point of suspension of the wire. The remainder of the cable is insulated by the air surrounding it.

The types of insulating material commonly used are:

Polyvinyl chloride (p.v.c)
Polychloroprene (p.c.p.)
Impregnated paper
Rubber: 60 °C
       85 °C
Ethylene-propylene (e.p.)
90 °C thermosetting
Mineral insulation
Other, types for special purposes.

Large cables may be single-core, twin-core, triple-core, or multi-core, in which form the cores are separately insulated and laid side by side with a slight continuous 'lay', packed with wormings and further insulated overall to form a circular shape. They may also be made as twin concentric or triple concentric. Smaller cables for lighting and power are laid up in flat form to give neatness of installation.

Polyvinyl chloride

This was originally introduced as a substitute for rubber. Its properties are generally similar to rubber, although it has a tendency to soften under moderate temperatures and to crack at low temperatures. It is practically impervious to chemical action. P.v.c.-insulated cables may be used where the combination of ambient temperature and temperature rise due to load does not exceed 65 °C. These cables are manufactured in the 600/1000 V range for installation purposes and in the range 1900/3300 V for electricity supply. P.v.c.-insulated cables are made up in many ways.

Vulcanized rubber (v.i.r.).

This material is fast losing ground owing to the common use of p.v.c. It is a preparation of pure rubber with a small amount of sulphur. It is impervious to water, flexible and of high resistivity. It has a fair mechanical strength dependent upon its degree of

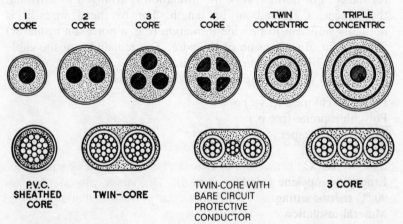

Fig. 3.1   Cable sections

vulcanization. It retains its properties for long periods in the absence of light and undue heat. The rubber is applied to the conductor during manufacture in two or more layers, or may be extruded. It is then vulcanized.

Impregnated paper

Paper-insulated cables are manufactured in voltage ranges from 600/1000 V to 19 000/33 000 V mainly for underground laying. The paper insulation is laid on to the conductor in the form of paper tapes laid helically to the desired thickness. The insulation, after being vacuum-dried, is impregnated under pressure with mineral oil or other suitable compound. Owing to the affinity of paper for moisture, the insulation is completely enclosed in a continuous lead or aluminium sheath extruded over the insulation. All cable ends are sealed by special oil-filled or compound-filled sealing boxes.

Rubber insulation

There are different types of rubber insulation available having different temperature limitations. The 60 °C rubber is used for general purposes, whilst the 85 °C rubber is used for connection within heating appliances or high ambient temperature environments.

Mineral insulation

Copper or aluminium conductors with compressed powdered mineral insulation enclosed in a copper or aluminium sheath (m.i.m.s.), may be used according to the termination used, up to an ambient temperature of 150 °C and to a much higher cable temperature. Specially designed arrangements are needed at all joints and terminations. These cables are made in the 600-V class (light duty) and the 1000-V class (heavy duty).

Mechanical protection

The mechanical protection of the smaller cables is dealt with in the section on 'Wiring systems' (Ch. 4). The larger cables, used for underground work and for large interior power installation, may be mechanically protected in various ways.

Unarmoured cables may be run without further protection than the lead sheathing. A further protection is one or two layers of compounded jute or hessian tape yarn laid over the lead sheath. Aluminium is used as an alternative to lead for sheathing cables.

Armoured cables include single-wire armouring (a single layer of galvanized iron wire laid spirally upon a bedding of jute or hessian), double-wire armouring (two layers of armouring), and double steel tape armouring (two layers of steel tape laid spirally over the bedding with an overall finish of jute and hessian).

Aluminium strip armouring is sometimes used as an alternative to wire armouring.

For underground cables, wire armouring is used where the ground is liable to subsidence, to prevent the cable from breaking, whereas steel tape armouring is needed where physical damage from stones or workmen's tools may be expected.

Flexible cords and cables

Flexible wires varying in size from 0.5 mm^2 (16/0.20) to 4 mm^2 (56/0.30) are called flexible cords. Flexible wires of larger sizes from 6 mm^2 (84/0.30) to 630 mm^2 (2257/0.60) are called flexible cables. Flexible cords and cables are so made of fine gauge wires as to be much more flexible than ordinary wiring cables and are used for such purposes as from ceiling rose to lampholder, or from socket-outlet to portable apparatus. In general they must not be used for fixed wiring.

The type of flexible cord to be used with a specific type of appliance or luminaire is specified in BS 3456 and BS 4533 respectively, and Appendix 10 in the 15th edition offers some notes on the application of flexible cords.

The types of flexible cables or flexible cords from which selection should be made are:

(a) Braided circular.
(b) Unkinkable.
(c) Circular sheathed.
(d) Flat twin-sheathed
(e) Parallel twin, only for the wiring of luminaires (where permitted by BS 4533).
(f) Twisted twin non-sheathed, only for the wiring of luminaires (where permitted by BS 4533).
(g) Braided circular twin- and three-core, insulated with glass fibre. Only where abrasion or unique flexing will not occur.

(h) Single-core p.v.c. insulated non-sheathed flexible cables to BS 6004.

There are several IEE wiring regulations that are applicable to flexible cords and cables and these are:

Reg. 523-3   Which requires that every flexible cord with an accessory, appliance or luminaire should be suitable for the temperature that may exist. Additional protection in the form of suitable insulating sleeves should be fitted over the cores.

Reg. 523-27   Where flexible cords may be subjected to mechanical damage, they should be of a type having rubber or p.v.c. sheathing. It may be necessary to use an approved type.

Reg. 523-28   Braided circular twin- and three-core flexible cords having glass-fibre insulation may only be used where the risk of abrasion or undue flexing is absent.

Reg. 523-29   Flexible cords should not be used as fixed wiring.

Reg. 523-30   Flexible cords are normally only used for connection to portable appliances or equipment. Within this regulation an electric cooker having a rating greater than 3 kW is not considered to be a portable appliance.

Reg. 523-31   All exposed flexible cable or flexible cord should be as short in length as possible to avoid mechanical damage (length is normally 1.5 m–2 m for many portable appliances). Flexibles will be connected to the fixed wiring by a suitable accessory or enclosure, and will require a suitable device offering overcurrent protection, isolation and switching.

Reg. 523-32   When a flexible cord supports or partly supports a luminaire, the maximum mass to be supported for a given size of flexible cord is

CSA of conductor (mm^2)	Maximum mass (kg)
0.5	2
0.75	3
1.0	5

Reg. 524-1   The colour identification of the protection conductor in a flexible cable or flexible cord is green and yellow and like the other cores should be identifiable throughout its length (Reg. 524-4).

Generally the colour brown will identify the phase core in a flexible cord, and blue the neutral. Some three- and four-core flexible cords used for three-phase purposes will have three brown cores which will require identifying at each end by suitable coloured or numbered means.

The termination of flexible cables or flexible cords should be undertaken with care and all mechanical clamps or compression type sockets must retain all the wires of the conductor or core (Reg. 527.1).

Reg. 553-10　Any coupler used to join one flexible to another should be non-reversible in order to maintain correct polarity, and should have provision for maintaining continuity of the protection conductor. When such a coupler is used the male plug should be fitted to the loadside of the circuit (Reg. 553-11).

It should be understood that selection of a flexible cable or flexible cord has to have the same consideration as that for fixed wiring cables. Such calculations that are necessary will be shown in examples, but in general terms one has to consider:

(a)　the current-carrying capacity;
(b)　protection against overcurrent;
(c)　the environment in which it has to operate.

As the 15th edition of the *Wiring Regulations* requires a more detailed consideration of cable sizing it is considered that any calculations required for the City and Guild of London Institutes Part II installation course will only relate to the principles. The Part II (c) course will require students to be able to determine a particular cable size, having applied all the appropriate correction factors.

Before any calculations are considered, however, the general requirements for the installation and protection of fixed wiring cables must be considered.

Non-flexible cables

Every non-flexible cable operating at low voltage should be selected, as appropriate, from the following list. Every type of cable having a cross-sectional area of 10 mm^2 or less must either have copper conductors or copperclad aluminium conductors. Sheathed cables (p.v.c. lead, HOFR, etc.) may, if intended to be suspended to span some distance above ground level, incorporate a catenary wire. This wire avoids any strain or stress being imposed upon the cable conductors or insulation.

Cable types

1. Non-armoured p.v.c. insulated (BS 6004, 6231, Type B or 6346).
2. Armoured p.v.c. insulated cables (BS 6346).
3. Split concentric copper conductor p.v.c. insulated cables (BS 4553).
4. Rubber insulated cables (BS 6007).
5. Impregnated-paper insulated cables, lead sheathed (BS 6480).
6. Armoured cable with thermosetting insulations (BS 5467).
7. Mineral-insulated cables (BS 6207 Part I or Part II) with, where appropriate, fittings to BS 6081.
8. Consac cables (BS 5593).
9. Cables approved under Regulation 12 of the Electricity Supply Regulations.

Table 9A within the Appendix of the IEE *Wiring Regulations* (15th edn) lists some fifteen typical methods of installing cables and conductors, giving each method a letter as a means of coding or designation.

For example:

Type A. Single-core and multicore cables enclosed in conduit.

Type B. Single-core and multicore cables enclosed in cable trunking.

Type C. Single-core and multicore cables enclosed in underground conduits, ducts.

Type E. Sheathed cables clipped direct to or lying on a non-metallic surface.

Type F. Sheathed cables on a cable tray, bunched and even closed.

These letter codes will appear in specifications indicating which type of cable is to be installed using a specific method.

Applicable regulations are as follows:

Reg. 413-2   The impedance of relevant circuits (cables) have to be co-ordinated with the characteristics of the protective devices (fuses or circuit breakers) and the earthing arrangements for the installation so that in the event of an earth fault, any voltage that may exist between exposed and extraneous conductive parts will be of such magnitude and duration so as not to cause danger. It is this regulation which necessitates the automatic disconnection of a faulty circuit within a specified time (0.4 sec. or 5 sec.) for a particular type of circuit.

In order to achieve this co-ordination the nominal current rating or setting of the protective device I_N is not less than the load current (Design Current). I_B and I_N does not exceed the lowest of the current-carrying capacitors I_Z of any particular circuits conductors. In addition, the current causing the protective device to operate (I_2) does not exceed 1.45 times the lowest of any of the current-carrying capacities (I_Z) of any of the conductors in a circuit. These conditions can be expressed as:

$$I_B \leqslant I_N \leqslant I_Z \text{ and } I_2 \leqslant 1.45 \ I_Z$$

Ref. 434-6 and Tables 54B, C, D and E give the values for the factors K which are used in the calculations to determine whether a cable will suffer damage due to temperature rise under short-circuit faults. The factor K is based upon the type of conductor material and insulation material in the cable.

Reg. 522-1 Every cable conductor size must be such that its current-carrying capacity (I_Z) is not less that the maximum sustained current that will flow through it under load conditions.

Reg. 522-3 Cables connected in parallel must be of the same type, c.s.a., length and disposition so that as far as it is possible all cables will carry equal current.

Reg. 522-4 Where bare conductors as used in busbar trunking systems are to be installed, then due account must be taken for their expansion and contraction, any joints or where main cables connect to them. This is normally the function of the manufacturer.

Reg. 522-5 Where cables are to be connected to bare conductors or busbars, the cable insulation and sheath should be suitable for the maximum operating temperature of the bare conductor or busbar (recommended maximum is 90 °C). Should there be any doubt then the insulation of the cable should be stripped back for some 150 mm from the tremination and this insulation replaced with suitable heat-resisting insulation.

Reg. 522-6 Concerns itself with cables that may be in contact with, or surrounded by, thermal insulation. If practical, such cable runs should be such to avoid contact with glass-fibre, expanded polystyrene, etc. However, if this is not possible then correction factors may have to be applied to the current-carrying capacity of the cable.

For a cable in contact on one side with thermally insulating material, the correction factor will be 0.75; for a cable completely surrounded by thermally insulating material the correction factor is 0.5. The designer of the installation would have to consider the type of installation and whether the cables are fully loaded, either intermittently or continually; e.g. in the case of a domestic installation, a lighting circuit in a roof void would never be operating at its maximum current rating. Hence the cable would not generate heat.

Reg. 522-7 Requires the metallic sheaths and/or non-magnetic armour of all single core cables in the same circuit to be bonded together at both ends of their route. Or, for conductors exceeding 50 mm^2, such bonding may be effected at one point in their run. This is to avoid the flow of eddy currents which may create sheath or armour to earth voltages or create corrosion.

Single core cables armoured with steel wire or tape must not be used on a.c. installations, and single core cables enclosed in ferrous (steel) conduit or trunking should be so arranged that all phases and the neutral are contained together. This is to prevent the magnetic effect being increased by and within the steel enclosure resulting in possibly severe heating.

Reg. 521-8 The size of every conductor (excluding those fed from extra low voltage (ELV) sources (see definition within Part 2 IEE *Wiring Regulations* (15th edn)) shall be such that, for circuits protected by devices not greater than 100 A, the voltage drop from the origin of that circuit to any other point in that circuit does not exceed 2.5 per cent of the nominal (supply) voltage when the design current is flowing.

Reg. 522-9 In general requires that the neutrals in polyphase circuits, in which imbalance is likely to occur, should be of the same size as the phase conductors. Certainly for discharge lighting circuits this must be so. Where it can be assured that serious imbalance is unlikely to be sustained, however, provision can be made to use a reduced neutral conductor.

Reg. 522-10   All conductors and cables shall have adequate strength and be so installed so as to withstand any electro-mechanical forces that may be created under service conditions, including short-circuit current.

Reg. 523-1   The type and current-carrying capacity of every conductor, cable, termination and joint must be suit-able for the highest operating temperature that is liable to occur in service. This requirement applies to transfer of heat from any appliance or luminaire to which the conductor or cable is connected.

Reg. 523-2   Cables installed in a heated floor or other heated parts of a building, and not actually heating cables, should be related to the normal operating temperature of the heated floor or part of building.

Reg. 523-3   As considered previously for flexible cords, when a cable is terminated within an accessory, appliance or luminaire, that cable must be of a suitable type or be provided with additional insulation compatible with the temperature that may exist within the enclosure.

Reg. 523-8   All metal sheaths, armours and fixings for cables when exposed to the weather must be of corrosion-resistant material or finish. Neither should they be in contact with any other metal such that electrolytic action may occur. It is for this reason that brass fixing screws should be used with mineral-cable fixing clips and saddles.

Reg. 523-9   Copperclad aluminium conductors should not be used where any terminations are exposed to wet conditions; for example, earthing conductors to externally sited earth rods (electrodes).

Reg. 523-10   Aluminium conductors should not be terminated into brass or other metal with a high copper content, unless the terminal is suitably plated (usually nickel plated) or other precautions are taken to maintain electrical continuity.

Readers should note that aluminium oxidizes very quickly when insulation is removed. In practice a thin smear of Densal grease, which is conducting, avoids oxidizing.

Reg. 523-11 The joints and terminations of cables insulated with impregnated paper must be protected against the entry of moisture by effective sealing. This is done either with a bitumastic compound or an expoxy resin.

Reg. 523-12 As in Reg. 523-11 the entry of moisture must be prevented when mineral insulated cables are terminated. Like impregnated paper, the mineral insulant magnesium oxide is hydroscopic. This means that it will absorb moisture and hence lose its insulating property. The insulating and moisture-proofing properties of the plastic compound used in mineral insulated (MI) cable terminations must be retained across the temperature ranges found in service.

Reg. 523-19 All cables should be protected against risk of mechanical damage. This is why it is necessary to encage some cables in a protective capping or conduit. This capping or conduit if metallic and if used solely for mechanical protection does not have to be earthed.

Reg. 523-20 Where cables are installed under floors or above ceilings, they should be run so that they will not be damaged by floor or ceiling fixings. Within the traditional timber floor construction this is achieved by drilling holes through the joist such that the cable is 50 mm from the top, or bottom as appropriate, of the joist.

Otherwise, cables not having earthed metallic sheaths should be enclosed in earthed metallic conduit or by equivalent mechanical protection to prevent damage due to nails, screws and other fixing devices.

Reg. 523-21 Where any cable passes through a hole in metalwork, precautions must be taken to prevent damage to the cable. This is normally done by using bushes, grommets or cable glands.

Reg. 523-22 Non-sheathed cables must be enclosed in conduit or trunking. Such cables must not be drawn into ducts formed by concrete or cement. The commonly used p.v.c. sheath is there to offer mechanical protection to the p.v.c. insulation.

Reg. 523-23 Cables buried directly in the ground must incorporate an armour or metal sheath or both. This armouring is to be protected against corrosion and current practice is to use a cable type having a p.v.c. sheath. All such buried cables should be marked with cable tiles or covers. Alternatively, a p.v.c. marking tape may be used. Whilst there is no minimum depth for such cables, they should be at a sufficient depth to avoid damage should the ground be disturbed.

Reg. 523-24   Follows on from Reg. 523-22 and requires cables that are to be installed in underground ducts to be sheathed or armoured, in order to resist any mechanical damage during the drawing in of cables.

Reg. 523-26   and Appendix II deal with the length and height of spans for cables between buildings. The requirement is that all such cables are at such a height to be out of reach of mechanical damage.

Reg. 523-35   Requires that cables and wiring systems installed in positions open to direct sunlight to be of a type to resist damage by the ultra-violet rays of the sun. Generally black is the best colour for any sheathing, although orange p.v.c. sheathed MI cable is commonly used. Orange being the identifying colour for electrical services.

Reg. 524-3   Requires all cable cores of fixed wiring systems to be identifiable at terminations and preferably throughout their length. Various methods are used according to the type of insulation. Such methods will involve coloured cores, tapes or sleeves or discs. With a numbered core method 1, 2 and 3 are used to indicate the phase conductors and zero (0) for the neutral.

Table 52A indicates the colour coding to be used for the full range of circuits employed.

Regulations 525-3 to 525-9 deal in some detail with the segregation of certain categories of circuits. From the definitions within Part II of the 15th edition it may be seen that a

Category 1   circuit is one, other than fire alarms or emergency lighting circuits, operating at low voltage and supplied directly from the mains supply.

Category 2   circuit is one, apart from the fire alarms and emergency lighting circuits, for telecommunications (radio, bells, and data transmission) supplied from a safety source (Reg. 411-3).

Category 3   circuit may be either a fire alarm circuit or emergency lighting.

As these categories of circuits impose restrictions on methods of unstalling circuit cables it is necessary to consider the main features of the relevant regulations.

Cables of category 1 circuits may not be drawn into the same conduit, duct or ducting as those of category 2 circuits, unless the latter cables are insulated to the same voltage grade as for the category 1 cables (Reg. 525-3).

Category 1 circuit cables may not be drawn into the same conduit, duct or ducting as category 3 circuits under any circumstances.

BS 5266 governs emergency lighting systems and further recommends that the emergency lighting circuit cables are segregated from cables of any other circuit (Reg. 525-4).

If a common channel or trunking contains category 1 and 2 circuits, then these circuits must be segregated by a partition. Unless as in 525-3, the category 2 circuit cables have the same grade of insulation as the category 1 circuits (Reg. 525-5).

This regulation is required to be more stringently complied with when category 3 circuits are contained in a compartmentalized trunking, all outlets and cross-out positions have to be effectively separate. However, should the category 3 circuits be wired in mineral-insulated cable, such partitions are not required (Reg. 525-6).

Control or outlet positions for category 1 and 2 circuits mounted on common boxes or switchplates as part of a conduit, ducting or trunking system must be effectively separated at such outlet positions by rigidly fixed screens or barriers (Reg. 525-7). BS 5266 also recommends such partitioning of emergency lighting circuits from any other circuits.

Where cores of category 1 and 2 circuits are contained within a common multicore cable, flexible cable or flexible cord, then the category 2 cores shall be insulated to the same grade as the category 1 cores. When such multicore cables terminate, then the segregation required by Reg. 525-7 will again apply.

Alternatively, the separated categories may be brought out into separate and distinctively marked terminal blocks (Reg. 525-8).

Cores of category 1 and 3 circuits may not be contained in a common multicore cable, flexible cable or flexible cord (Reg. 525-9).

Regulation 525-12 stipulates that only lift installation cables may be run in a lift shaft.

All cables and conductors must be supported so that they will not be subjected to undue mechanical strain. Also there should be no strain on terminations of conductors (Reg. 529-1). It is for this reason that Appendix II offers guidance on methods of supports. Being rather detailed, it is necessary to refer to this appendix.

Regulation 529-3 requires that the internal radius of any bend in a non-flexible cable shall be such that no damage will be done to the insulation; in other words, the bends must not be too sharp. Table 52C in the 15th edition gives all minimum internal radii of bends for a range of cable types and reference to this table is required.

Regulation 529-7 exists in order that the number of cables drawn into or laid in an enclosure, conduit or trunking, will be such that no damage will be caused to the cable or enclosure during installation.

Appendix 12 in the 15th edition describes a method by which it is possible to calculate the size of conduit or trunking to accommodate cables of the same or differing sizes.

In the method employed, each cable size is allocated a factor. The sum of all factors for the various cables to be enclosed is compared against the factor listed for a given size of conduit or trunking. The correct size of conduit or trunking is one whose factor is equal to or just greater than the total cable factor. Again such is the detail within Appendix 12 and the six tables (12A–12F) that reference should be made to the appendix.

Cable selection

In order to select the correct cable size for a particular set of conditions, a number of stages or steps have to be undertaken. These will necessitate reference to tabulated data within the 15th edition.

Should a fault occur in a circuit or appliance connected to that circuit it is likely that current will flow to earth, either through any metallic sheathing of a cable, steel conduit or trunking, or via the casing of the appliance. The value of current flowing to earth is dependent upon the effectiveness of the connections to earth within the installation, together with the supply network. Should there be any loose connections or poor joints, then the opposition to the earth fault current flow will be increased. This opposition to current flow is called the impedance. The term impedance rather than resistance is used because alternating current rather than direct current is involved.

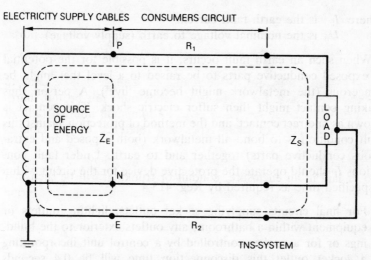

Fig. 3.2 (CITB) The impedance to the earth fault current $Z_s = Z_E + R_1 + R_2$

For the simple circuit shown in Fig. 3.2 the total impedance to the earth fault current, known as Z_s, is the impedance of the system, the system being the supply network (supply source) comprised of substation transformer secondary winding and the distribution cables into the service (origin of supply), together with the phase conductors impedance (R_1) and the impedance of the protective conductor (R_2). For these conductors, if they are 35 mm^2 or less, any inductive effects may be ignored and thus their impedances equal their resistances.

For the radial circuit shown in Fig. 3.2, and as indicated in Appendix 8 of the 15th edition,

$$Z_s = Z_E + R_1 + R_2$$

Where Z_s = Total impedence of the system for the earth fault current

$\quad\quad\quad Z_E$ = Impedance of supply network or source

$\quad\quad\quad R_1$ = Phase conductor impedance

$\quad\quad\quad R_2$ = Protective conductor impedance

Having established Z_s it is possible to calculate the earth fault current I_F from

$$I_F = \frac{U_o}{Z_s}$$

Where I_F is the earth fault current

U_o is the nominal voltage to earth (supply voltage)

When such an earth fault occurs, it is possible for the potential of exposed conductive parts to be raised to a level that might be dangerous (the metalwork might become 'live'). A person thus making contact might then suffer electric shock. This contact is known as 'indirect contact' and the method of protection against this fault condition is to bond all metalwork (both exposed and extraneous conductive parts) together and to earth. Under fault conditions I_F should operate the protective device for the circuit within a specified time as required by Reg. 413-4

1. For final circuits supplying socket outlets, lighting outlets or equipment within a bathroom, any outlets exterior to the buildings or for a cooker controlled by a control unit incorporating a socket outlet this disconnection time will be 0.4 seconds maximum.
2. For final circuits supplying fixed equipment (except as in 1) disconnection time will be 5 seconds maximum.

This method of protection against indirect contact is known as EEBAD (Earthed Equipotential Bonding and Automatic Disconnection). It is the basis for all cable calculations now as the size and length of run for circuit conductor will be fixed by the values of R_1 and R_2 related to Z_E (this value being obtained from the supply authorities) and the rating and type of protective device used.

The stages in selecting a cable size and verifying that it meets all requirements are to verify that:

1. Volt drop constraint is within permitted limits.
2. Shock protection is provided.
3. Thermal constraints are met.

As stated earlier in this book, students following Part II–236 courses would be expected to work to this level, but Part II students would not have to deal with calculations related to shock protection and thermal constraints.

Sequence of operations

The sequence of operations involved in meeting these stages are as follows:

1. Calculate the design current I_B for the circuit.

2. Select appropriate type and rating of protective device from Table 41A1 or Table 41A2 to give I_N.
3. Apply correction factors that may be appropriate, to I_N as divisors in order to establish the current-carrying capacity of the cable I_Z.
4. Select cable from the Table 9D1 to 9N1 within Appendix 9, according to conductor material and cable type.
5. Verify that circuit voltage drop is within permitted limits.
6. Determine that shock protection is afforded.
7. Ensure that the thermal constraints are met.

EXAMPLE

A circuit supplying a 240 V–12 kW domestic electric cooker is protected by a BS 3036 rewireable fuse. The circuit cable has p.v.s. insulation and sheathing contains a c.p.c. and the circuit route is 15 m in length. A 13 A socket outlet is incorporated in the cooker control unit and the value for Z_E is 0.35 Ω.

If the cable runs for most of its length in the roof void and is in contact on one side with glass fibre insulation, determine the minimum size of the cable to be installed and yet meet all requirements.

Maximum full load current

(i) $I = \dfrac{W}{V} = \dfrac{1200}{200} = 50A$

(ii) *Design current* (applying diversity, Table 4A/4B)
= 10 A + 30% of (40 A) + 5 A
= 10 + 12 + 5
∴ $I_B = 27$ A

(iii) From Table 41A1, as cooker control unit has a socket outlet:
I_N = rating of fuse   (30 A (in (c)))
And under this value Z_s is given as 1.1 Ω

(iv) *Current capacity of cable* $I_Z = \dfrac{I_N}{C_1 \times C_2 \times C_3 \times C_4}$

Where C_1 = Correction factor for ambient temperature
C_2 = Correction factor for grouping of cables
C_3 = Correction factor for thermal insulation
C_4 = Correction factor if BS 3036 fuse is used

Therefore in this instance

$$I_Z = \frac{I_N}{C_3 \times C_4}$$

as only contact with thermal insulation (on one side) and use of rewireable fuse has to be considered

$$= \frac{30}{0.7 \times 0.725}$$

$$= 59.1A$$

(v) From Table 9D2 (twin and multicore p.v.c. insulated cables), Column 6
10 mm^2 cable has a rating of 64A

(vi) *Volt drop (VD) for this circuit*
As maximum permitted VD is

$$\frac{2.5}{100} \times \frac{240}{1} = 6 \text{ V}$$

$$\text{Actual VD} = \frac{I_B \times \text{length of run} \times \text{tabulated V/A/M}}{1000}$$

$$= \frac{27 \times 15 \times 4.2 \text{ (column 7, Table 9D2)}}{1000}$$

$$= 1.7 \text{ volts}$$

Thus the volt drop is satisfactory, being below the maximum permitted

(vii) *Shock protection*
Z_s allowed for 30 A – BS 3036 fuse from Table 41A1 is 1.1 Ω
∴ $Z_E + R_1 + R_2$ should not exceed this value,
From Tables 17A and 17B in Appendix 17
Resistance per metre for 6 mm^2 phase conductor and 4 mm^2 c.p.c.
= 6.44 Ω/m × multiplier of 1.38
* Resistance for 15 metres = 0.00644 × 1.38 × 15
$$= 0.133$$
Therefore $R_1 + R_2$ = 0.133
$$Z_s = Z_E + R_1 + R_2$$
$$= 0.35 + 0.133$$
$$= 0.483$$
This value for Z_s is well within the limit set by Table 41A1.

(viii) Check for compliance with thermal constraint (Reg. 543-2).
Fault current that may flow

$$I_\text{F} = \frac{U_\text{o}}{Z_\text{s}}$$
$$= \frac{240}{0.483}$$
$$= 497 \text{ A}$$

From the time current characteristic (graph) Fig. 11 in Appendix 8, the time taken for a 30 A – BS 3036 to operate under this fault condition would be 0.1 second. In order to ensure the size of the c.p.c. is adequate to meet the thermal constraint the formula given on Reg. 543-2 should be used.

$$S \ne \frac{I^2 t}{K}$$

where S is the c.p.c. size

I is the fault current I_c

t is operating time of the disconnecting device in seconds

K is the factor dependent upon conductor and insulation material and obtained from Table 54C

$$\therefore S = \frac{494.8^2 \times 0.1}{115}$$
$$= 1.36 \text{ mm}^2$$

Nearest standard size of c.p.c. is 1.5 mm (Reg. 543-2)
∴ 4 mm² c.p.c. meets the thermal constraints.
So cable to install is 10 mm² twin with 4 mm² c.p.c.

This was an extended example intended to show the various stages of working and students of CGLI 236 courses, Part II would not have had to complete all stages. However, those having to specify or select cable sizes for a particular circuit or installation would invariably meet a range of other situations.

Another way of calculating the voltage drop in a cable involves using Ohm's law. By considering formula, $V = R \times I$, it will be seen that the voltage drop in the conductors of a cable is the product of the current in the conductors and the resistance of the conductors. This applies to direct current circuits, and to those alternating current circuits where the effect of the inductance and capacitance of the cable can be neglected.

The resistance of a conductor is calculated from the formula

$$R = \frac{\rho l}{a},$$

where R is the resistance of the conductor in ohms, ρ is the resistivity of conductor material in microhm millimetres, l is the length of the conductor in millimetres and a is the cross-sectional area of the conductor in square millimetres.

When calculating the voltage drop in a 2-core cable, the lengths of both lead and return conductors are taken into account, and the value for l is twice the cable length.

EXAMPLE

A 2-core copper cable supplies current to a 240-V single-phase load of 18 kW at 0.78 power factor. The cable is 40 metres long, and each conductor has a cross-sectional area of 35 mm^2. Calculate:
   (i) the voltage drop in the cable at the load, ignoring the reactance of the cable,
   (ii) the power lost in the cable.
The resistivity of copper may be taken as 17.5 $\mu\Omega$ mm.   (C.G.L.I.)

(i)   Load current $= \dfrac{\text{W}}{\text{V} \times \text{p.f.}} = \dfrac{18\ 000}{240 \times 0.78} = 96.1$ A

Conductor resistance (lead and return)

$$= \frac{\rho l}{a} = \frac{17.5}{10^6} \times \frac{40 \times 10^3}{35} \times 2 = 0.04\ \Omega$$

Therefore, total volt drop $= RI = 0.04 \times 96.1 = 3.84$ V

(ii)   Power lost in cable $=$ volt drop $\times$ current $= 3.84 \times 96.1$
$= 369$ W

Alternatively, power lost $= RI^2 = 0.04 \times 96.1^2 = 369$ W

Joints and jointing

I.E.E Reg. 527-1 requires that joints in cable conductors and bare conductors shall be electrically and mechanically sound. Joints in non-flexible cables are to be accessible for inspection except when buried underground or enclosed in non-combustible building materials. Joints in these non-flexible cables are to be made by soldering, welding or brazing, or by the use of mechanical clamps, except that mechanical clamps shall not be used for inaccessible joints.

In the modern wiring systems described later, mechanical joints are mostly used, either as special joint boxes with connector terminals integral with the box.

In the looping-in system using p.v.c.-insulated cables, the joints are twisted dry joints clamped in the terminals of switches, ceiling roses, etc. Soldered joints are often used, however, particularly for jointing the larger size cables, for which reason a knowledge of jointing by soldering is essential to the electrical wireman.

Accessibility

The requirement that all joints shall be readily accessible means that joints of any type must not be pulled into conduit tubing or hidden in partitions, but must be enclosed in conveniently placed inspection-type joint boxes.

Soundness

A joint which is mechanically sound must not pull out under the normal strains to which it may be subject during installation or in its working life. To be electrically sound, the joint must have at least as low a resistance as an equal length of the conductor itself, nor must it overheat when carrying current. To effect this low resistance, the cables to be joined must be in intimate contact over the whole of the contact area, and if the joint is soldered the solder must be so used as to *hold* the copper wires in close contact. In other words, a soldered joint should be mechanically and electrically sound *before* soldering.

Soldered joints

The solder used for electrical joints in copper is a soft solder ('tinman's'), an alloy of equal parts of tin and lead. Its melting point is about 200 °C, roughly one-fifth of the melting temperature of copper, and still much above the usual temperature limits at which cables may be worked. The only jointing flux which should be used for electrical joints is resin. Killed spirits should on no account be used.

For jointing aluminium, special fluxes have been developed. Aluminium oxidizes very rapidly, and in the past much difficulty has been experienced in the making of a good joint. Nowadays, special solders and fluxes are available which inhibit the oxide film and allow a satisfactory joint to be made.

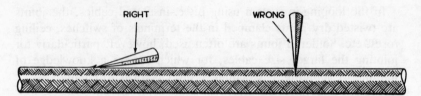

Fig. 3.3   Removing insulation

Preparing the cable

The instructions which follow apply to p.v.c. cables, v.i.r. cables and the like. The insulation is removed by cutting with a sharp knife. The blade should be held at a narrow angle to the wire and used with a paring motion (Fig. 3.3). Cutting the insulation at right-angles to the wire leads to nicking the wire, increasing its resistance at the point damaged, and to possible breakage of the strand.

The copper wires in cables other than v.i.r. are not tinned, and each strand must be properly tinned before proceeding with the joint. When the cable is properly trimmed, cleaned, and tinned, the joint may be made, soldered, and insulated.

Soldering methods

The correct method of soldering is to heat the wire to be soldered to such a temperature that it will melt solder put into contact with it. For instance, soldering cannot be satisfactorily accomplished by simply dropping blobs of melted solder on to the cold wire. The heating is done either by means of a soldering-iron for small cables up to 2.5 mm², or by the use of a gas torch or solder-pot and ladle for larger cables.

Use of the gas torch

When soldering larger cables from 4 mm² upwards, the amount of heat held by the soldering-iron is insufficient to heat the joint. Direct heating of the joint by means of the gas torch is necessary. Again, proper preparation, cleaning, and tinning of the wires will facilitate quick operation. During the heating process flux is frequently applied. When the joint is at a suitable temperature the solder is applied direct to the hot surface of the joint, melting and running over and through it. The joint should be twisted if possible during the process to ensure that the solder is not adhering only to the top surface. The danger of burning the insulation of the cable is overcome by previously wrapping the insulation with asbestos string or in some similar manner. The remaining procedure is as in the last paragraph.

Using the metal-pot and ladle

For the larger cables many jointers prefer to use the metal-pot.The solder is heated in the metal-pot to the required temperature, over a fire or on a special 'roarer' blowlamp. The joint is covered with the flux and the solder is poured onto it from one ladle, that which runs off being caught in a second ladle. The hot solder itself heats the joint sufficiently, and a few consecutive pourings will suffice to solder the joint completely, making a sound joint. Finish off as before.

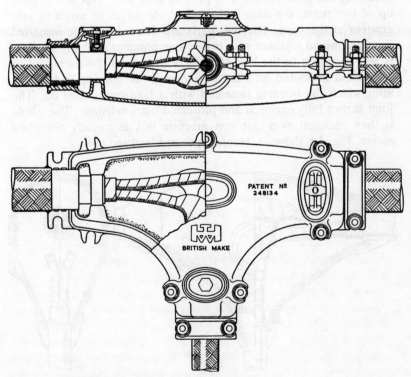

Fig. 3.4 'Henley' tee box (W. T. Henley & Co Ltd)

Jointing paper-insulated cables

Paper-insulated cables as used for underground distribution are entirely enclosed in an extruded lead sheath. All joints also must be entirely enclosed. The conductors are jointed as already described, by the use of the metal-pot and ladle. Cotton tapes boiled in insulating oil are used as the primary insulation. The whole joint is then enclosed in a cast iron junction box consisting of two parts, the bottom half and the top half. The box is fitted with clamps for the armouring (if any), and with glands to fit the leads sheaths. When the box is bolted together, mechanical and electrical continuity is thus assured. A bituminous insulating compound is poured hot into the box through the filling hole. The compound sets when cold, holding the cores in place and entirely excluding moisture. The box is then 'topped up' with compound, and the filling plate is bolted down. Figure 3.4 shows a 'Henley' tee box.

Many engineers take further precautions against moisture by enclosing the joint proper in a pressed lead box. This box is built up of two parts, the base and the top, the edges of which fit into grooves in the base. The two parts are fitted round the insulated joint and wiped together with two-to-one plumber's solder. The box is also wiped on to the lead sheaths of the cables. Insulating oil or compound is poured into the box through an opening in the top, after which the opening is sealed with a lead cap, wiped on. The joint is thus fully enclosed and protected from moisture. The whole is then enclosed in a cast-iron junction box as already described above. Fig. 3.5 shows a 'Henley' pressed-lead tee box.

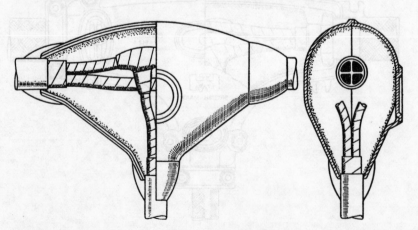

Fig. 3.5   Pressed-lead tee box (W. T. Henley & Co Ltd)

4 Wiring systems

A wiring system consists of the conductor, its insulation, its mechanical protection, and the various accessories, such as joint boxes, etc. The systems are named mainly in terms of the mechanical protection used.

In considering the use of any particular wiring system, it should be realized that no system can be the ideal one under all conditions.

A number of points must be considered, e.g.: neatness of the finished job; the durability of the installation; future extensions and alterations; the time required to do the work; damage to the fabric of the building by cutting away ; special conditions to be withstood, such as fumes, dampness, etc.; and the total cost of the job. A surface system will normally necessitate much less cutting away than a hidden system.

The various systems used for lighting and small power are:

Bare conductor wiring
Steel conduit:
(a) Slip-joint conduit with grip fittings (ELV installations only)
(b) Screwed conduit
Non-metallic conduit
Rubber-sheathed (t.r.s.)
p.v.c.-sheathed
Earthed-concentric
Mineral-insulated metal-sheathed (m.i.m.s.)
Special systems for different conditions

Bare conductors

Lightly insulated or bare conductors may be used for such purposes as earthing connections, rising mains and busbar systems, collector wires for cranes, etc. They should not be used where flammable or explosive dust, vapour, or gas is present or where explosive materials are handled or stored.

Bare conductors used for rising mains or busbars should be installed only in places inaccessible to unauthorized persons, and be supported by insulators so as to be free to expand and contract with changes of temperature.

When bare conductors are used as collector wires, screens or barriers must be erected to prevent accidental contact, and warning notices should be affixed. Conductors passing through walls, floors, partitions or ceilings shall pass through directly and be protected by incombustible insulating material, or earthed metal trunking.

Steel conduit systems

Annealed mild steel tubing is very widely used for enclosing v.i.r.- or p.v.c.-insulated cable or any other insulated cable.

The conduit is specially annealed so that it may readily be bent or set to any angle without breaking, splitting, or kinking.

B.S. specifications govern the manufacture of the classes of conduit given.

Light-gauge conduit (plain)

(*a*) Close joint: The tube is formed from a strip of metal bent into shape with the edges butted together without a mechanical joint.

(*b*) Brazed or welded joint: This is similar to close joint, but the seam is mechanically joined by brazing or welding.

(*c*) Seamless or solid-drawn: This is produced by cold-drawing from the bar.

Heavy-gauge conduit (screwed)

(*a*) Welded joint

(*b*) Solid-drawn

In each case both ends of each length of conduit are screwed electric thread.

Standard sizes

Common standard sizes of conduit are 16 mm ($\frac{5}{8}$ in), 20 mm ($\frac{3}{4}$ in), 25 mm (1 in) and 32 mm ($1\frac{1}{4}$ in), external diameters.

Enamel finish

The usual black external and internal finish for conduit and conduit accessories is a flexible stove enamel, which should not chip or flake in use. Light enamel finish is also available.

Galvanizing

To prevent the formation of rust when conduits are installed in exposed places, they may be specially treated externally and internally during manufacture with an effective covering of zinc spelter.

Sherardizing

This is a special process by which zinc alloy is embedded in the surface of the metal, making it rust-resistant.

Screwed conduit

Screwed conduit with screwed metal junction boxes and outlet boxes is generally used for the better-class installations. The conduit is supplied in lengths of from 3 m to 4 m, with both ends screwed. Special lengths may be supplied, and for repetition work time and material may be saved by ordering the conduit in bundles of fixed lengths. Long lengths should be used where possible, as short lengths invoel extra screwing and the insertion of extra couplings. The various junction boxes and outlet boxes for single conduits are internally screwed, and in the best class are fitted with shoulders to prevent the conduit from protruding into the box, and to obviate the need of bushes.

Bending conduits

Bending block.   It is often necessary to set or bend the conduit to enable it to pass over an obstruction or to turn a corner Fig. 4.1. A good electrician is able to make very neat sets, either by means of a bending machine or by using a simple bending block. The sets are made cold except in the larger sizes of conduit, when the tubing needs to be heated.

The bending block may be made from a piece of hardwood, say, 1.25 m × 150 mm × 50 mm. Holes slightly larger than the conduit

Fig. 4.1   Passing obstruction

are bored through the block from front to back and the openings are chamfered off. To bend a piece of conduit satisfactorily requires practice to prevent kinking. The conduit to be bent is inserted in the hole and hand pressure is brought to bear to bend the conduit slightly (Fig. 4.2). The conduit is then moved through the hole a very short distance, and the process of bending is repeated. By this method a gradual and neat bend is formed. A fierce pressure will kink the conduit and render it useless. A sharp bend is much more difficult to form than a longer bend.

Bending machines.   Bending machines are now used very widely especially for the larger sizes of conduit (see Fig. 4.2).

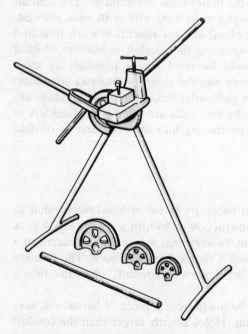

Fig. 4.2   Conduit bending machine

These completely portable outfits require no fixing or bolting down and are supplied with serrated jaw pipe vice for cutting and screwing purposes. They are fitted with a safety pin to secure the Bending Lever in an upright position when removing the bend from machine. Fitted with bosses to retain Lever in position when changing Formers and with positive and adjustable degree of bend quadrant to facilitate repetition bending. These benders are supplied as standard with Formers only to bend screwing size conduit.

When short-radius Pot Floor Formers are used it is also necessary to use Guides and a Plain Roller.

Radius of standard formers		Metric BS 4568			
Size	.. ..	16 mm	20 mm	25 mm	32 mm
Radius	.. ..	45 mm*	64 mm*	89 mm	102 mm

* All radii measured to inside of bend.

Threading conduit

The cutting of the screw thread is achieved by using stocks and dies of the appropriate size. The length of screw thread should be sufficiently long to fit half-way into the couplings and fully into the fittings. After screwing, the end of the tube must be properly reamered to clean off any sharp edges which will damage the cable insulation, and the oil used to lubricate the dies must be wiped and cleaned off. Before cutting and screwing the lengths, careful measurement is necessary, so that the switch and ceiling rose outlets fit in their exact positions.

Fixing the conduit

Conduit is fixed to bare brick walls by means of crampets, but on finished surfaces such as plastered walls, enamelled saddles or clips are preferred as making a neater job. Figure 4.3 shows a crampet, a clip, and various types of saddles.

The saddles and clips are screwed to plugs fitted into drilled holes in the wall surfaces. Large wooden plugs made on the job are unsuitable, as they dry out and become loose later. Proprietary plugs, such as Rawlplugs made for various sizes of screw, are very satisfactory and time-saving. The saddles are screwed directly to wooden surfaces and wooden joists. For surface work, spacing

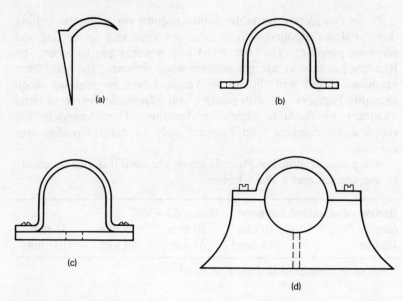

Fig. 4.3   Conduit fixings

saddles are often used to hold the conduit away from the wall and to facilitate wall decoration. Where running along or across steel girders, girder clips are used.

Conduit fittings should be screwed together very tightly, as a loose connection involves loss of electrical continuity.

All screwed joints should be painted, after erection, with a good lead or aluminium paint, which should also be applied to any part of the conduit where the enamel has been damaged.

A conduit run should terminate in a screwed brass bush if a terminal box is not used.

Running coupler

Sometimes it is impossible to screw the various parts of the conduit run together consecutively, and a running coupler is used to connect together two pieces of screwed conduit.

If it is not desired to use a manufactured running coupler, a running coupler may be made on the job, as shown in Fig. 4.4. One conduit A is normally threaded half the length of an ordinary coupler whilst the other conduit B is screwed sufficiently to accommodate the coupler C and a lock-nut D. The tubes are butted

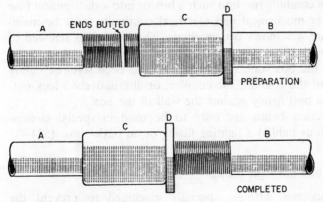

Fig. 4.4 Running coupler

together, and the coupler C is screwed tightly on to tube A, so that both tubes are held together. The lock-nut D is then screwed and tightened to butt against the coupler C and prevent movement.

Conduit fittings

Many different kinds of conduit fittings are available; the full range can be found in the manufacturers' catalogues. They include screwed elbows, bends and tees, non-inspection, inspection, and split types. Junction boxes, circular, oblong, or square.

A square 'adaptable' box is most useful when a number of conduits running together change direction. Figure 4.5 shows such

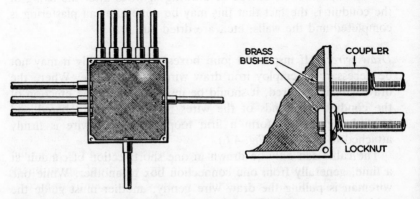

Fig. 4.5 Mechanical and electrical continuity at conduit box

a box. When conduits run into such a box or into a distribution fuse board, proper mechanical and electrical continuity must be maintained. Figure 4.5 shows two methods. The conduit is screwed of sufficient length to accommodate a coupler, a brass bush, and the thickness of the box wall. After the bush has been screwed tightly on the end of the conduit, the coupler, or alternatively a lock-nut, is screwed to butt firmly against the wall of the box.

Non-inspection bends are only to be used in special circumstances, such as behind a lighting fitting or an outlet box. (524-4)

Water-tight and gas-tight fittings

Conduit inspection fittings, specially machined to prevent the ingress of water or gas, are manufactured. The covers may be fitted direct or with gaskets.

Flexible conduits

Flexible metal conduit may be used in special situations such as the connections between motor starter and motor. Brass adaptors connecting flexible conduit to fixed screwed conduit are necessary. Flexible conduit must not be used as the c.p.c. conductor, and therefore a separate c.p.c. is necessary.

Drawing in the wires

The conduits of each circuit shall be erected complete before the cables are drawn in.

One important advantage of drawing in wires after the fixing of the conduit is the fact that this may be done after all plastering is completed and the walls, etc., are dried out.

Draw wires. If inspection joint boxes are used freely it may not be necessary to employ iron draw wires or steel tape. Where the draw wire is required, it should be inserted during the erection of the conduit. The ends of the wires to be drawn are bared and twisted together to form a firm loop. The draw wire is firmly attached to the loop (Fig. 4.6).

The cable can only be drawn in one short section of conduit at a time, generally from one connection box to another. While one wireman is pulling the draw wire gently, another must guide the cables into the tube. The cables must not twist round each other

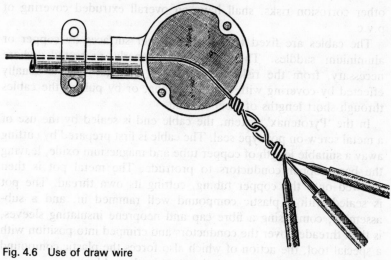

Fig. 4.6   Use of draw wire

but must be parallel throughout the run. If the cables twist, not only are they more difficult to pull in, but it will be impossible to withdraw any one if it be necessary at a later date. The wireman handling the cables may be able to guide up to three or four cables, but for a larger number some form of wooden guide block is necessary. This can be made on the spot, and may be a block of wood with the requisite number of holes through which the wires thread, or a circular block with slots cut in the rim.

Mineral-insulated metal-sheathed system

This type of cable (m.i.m.s.) is now extensively used because of its special qualities and advantages. The cable consists of a pliable copper or aluminium tube containing copper or aluminium conductors firmly positioned in highly compressed magnesium oxide powder. The cable may be used at much higher temperatures than rubber-insulated or p.v.c. cable, and is thus suitable for installations in boiler houses, heat-treatment shops and the like. The cable may also be used successfully under conditions of humidity and moisture.

The general Regulations are as for other types of metal-sheathed cables, although the cable ends must in every case be protected from moisture by sealing. M.i.m.s. cables, where installed in damp situations, or in concrete ducts, or where exposed to weather or other corrosion risks, shall have an overall extruded covering of p.v.c.

The cables are fixed to walls and other surfaces by copper or aluminium saddles. They must be properly protected, where necessary, from the risk of mechanical damage. This is usually effected by covering with steel sheathing, or by pulling the cables through short lengths of steel conduit.

In the 'Pyrotenax' system, the cable end is sealed by the use of a metal screw-on pot-type seal. The cable is first prepared by cutting away a suitable length of copper tube and magnesium oxide, leaving the bare copper conductors to protrude. The metal pot is then screwed onto the copper tubing, cutting its own thread. The pot is sealed with a plastic compound well rammed in, and a sub-assembly, comprising a fibre cap and neoprene insulating sleeves, is then threaded over the conductors and crimped into position with a special tool, the action of which also forces the plastic compound hard down onto the magnesia insulant in the cable.

Normally the cables enter into metal switch boxes, distribution boxes, etc., by special glands which screw into the boxes and hold

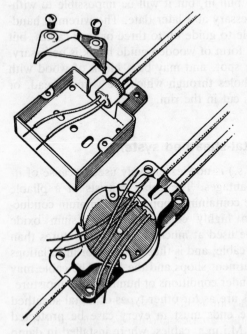

Fig. 4.7   Plaster depth box (BICC)

the sealed ends of the cables in secure grip. In special conditions flameproof glands may be used.

Figure 4.7 shows a type of box which employs a cable clamping arrangement which grips the cable sheath and allows the gland to be dispensed with.

Prefabricated wiring layouts

The demand for increased productivity in the building industry is being met in part by the factory production of standard building units which may be quickly and efficiently fixed together on site to form a complete building.

This leads to the production of wiring units of cables cut to length, assembled together and fitted into properly designed channels and ducts in the building units during factory assembly, or during site assembly.

Further uses of prefabricated wiring units can be considered, particularly in the case of building estates, where all the dwellings are alike in size and shape, or where similar rooms, e.g. kitchens and bathrooms, are built in different combinations to vary the overall sizes and shapes of the completed houses. There is obviously much scope for thought and ingenuity by architect, builder and wiring contractor, whereby economy in time and material can be achieved. Packaged units for various purposes are being manufactured.

Rising mains

A satisfactory system of rising mains consists of bare copper conductors mounted on porcelain insulators, the whole enclosed in metal trunking. Fused sub-circuit tapping boxes can be fitted to the trunking at any desired position, at each floor level in particular, so that all flammable insulating material is excluded from the interior of the trunking. A section of such a system is shown in Fig. 4.8.

Underfloor ducting

In many cases the final layout is not known when the electrical system is first installed in the carcase building. For general lighting from the ceiling, a conduit grid may be laid in position on the floor

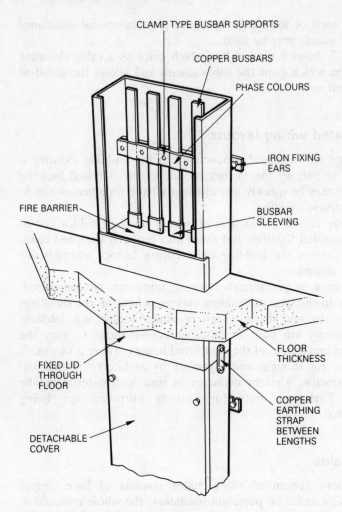

CLAMP TYPE BUSBAR SUPPORTS

COPPER BUSBARS

PHASE COLOURS

IRON FIXING EARS

BUSBAR SLEEVING

FIRE BARRIER

FLOOR THICKNESS

FIXED LID THROUGH FLOOR

COPPER EARTHING STRAP BETWEEN LENGTHS

DETACHABLE COVER

Fig. 4.8 Clamp type busbar supports

with a number of outlets to the ceiling below, or the grid may be fixed to the ceiling before the final cement screed is applied. The wiring is drawn in later to suit the lighting requirements. To give flexibility in the provision of outlets for all services, underfloor duct systems have been developed.

They may be divided into two kinds, providing either metallic or non-metallic protection to the wiring. One proprietary make of each kind is described below. In the general arrangements of each type a grid of ducting is laid on top of the solid cement floor over the whole floor area. The arrangement of the grid depends upon the use of the building. Where the layout in terms of rooms, corridors, etc., is known, the grid can be laid to a predetermined pattern, to provide the electrical services exactly where they are wanted. Where the layout is not known beforehand, a grid must be laid which will reasonably cover the area.

5 Wiring accessories

This section cannot pretend to give a comprehensive description of the very large variety of wiring accessories available for installation work. The illustrated catalogues of manufacturers will give much information, and should be available for students' reference in all college electrical installation workshops.

Lampholders

These are designed for quick removal and replacement of the lamp, and yet they must hold the lamp in firm metallic contact to prevent overheating. There are three main sizes of lampholder: the Bayonet-cap (B.C.), the medium Edison Screw (E.S.) and the Goliath Screw (G.E.S.). There are other variations such as the three-slot B.C. for the smallest discharge lamps. For ordinary tungsten filament lamps up to 150 W the lamp caps and thus the lampholders are B.C., up to 200 W the caps are E.S., and above 200 W they are G.E.S. In every case where a lamp is to be installed, the appropriate size and type of holder must be fitted. Lampholders may be either the insulated type of bakelite or the brass type with porcelain interior.

B.C. lampholders should have solid plungers separately sprung by rust-proof steel plunger springs, and in the insulated type a metal insert to reinforce the area around the bayonet sockets. An efficient cord grip is also necessary when the lamp is to be suspended from flexible cord. Figure 5.1 shows a range of insulated lampholders. The first type shown is a cord-grip lampholder partly sectioned. The section shows clearly the flexible wire securely fixed to the spring plunger, and also the method of gripping the flexible cord. The other types shown are a Home Office holder with shrouded skirt as installed in bathrooms, a pushbar switch holder, and a holder suitable for mounting directly on the ceiling. Figure 5.2 shows a range of lampholders. A Goliath screw lampholder is shown

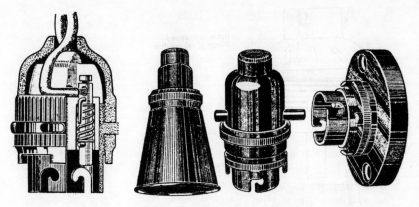

Fig. 5.1 Insulated lampholders (J. H. Tucker & Co Ltd)

Fig. 5.2

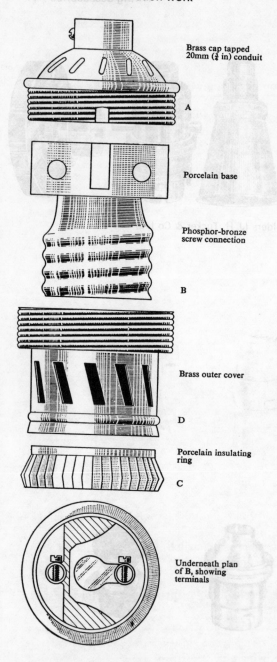

Brass cap tapped
20mm (¾ in) conduit

A

Porcelain base

Phosphor-bronze
screw connection

B

Brass outer cover

D

Porcelain insulating
ring

C

Underneath plan
of B, showing
terminals

Fig. 5.3 Goliath screw lampholder

exploded in Fig. 5.3. The outer contact is a screwed metal cylinder, and the inner contact is a metal stud. IEE Regulation 553-18 states that where centre-contact bayonet or Edison-type screw lamp-holders are connected to a source of supply having an earthed neutral conductor, the outer or screwed contact shall be connected to that conductor.

In wiring lampholders care must be taken in baring the flexible wire. The stranded wires must be well twisted together and should not be allowed to splay, as a loose single strand may touch either the metal frame of the holder or the opposite terminal.

Ceiling roses

Figure 5.4 shows a modern form of moulded ceiling rose which includes the earth terminal and a terminal for looping the phase conductor.

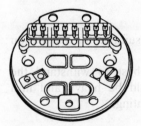

Fig. 5.4   Ceiling rose

Alternating current switches

Alternating current can efficiently be interrupted by the opening of a very small gap between silver contacts, and most switches are now of this type, the so-called 'microgap' switch. One of a range of 250-V 5-A switches is illustrated in Figure 5.5. The gap is set to about 6 mm when open.

Many varieties of switch are available: surface type, semi-recessed type, sunk type with switch plate, special types for fitting in conduit boxes, etc. The reader is advised to study a switch manufacturer's catalogue in this connection. Special note should be taken of the single-cord ceiling switch, which has a reciprocating action, whereby one pull of the cord puts the switch on, and the following pull puts

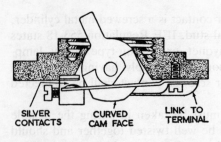

SILVER CONTACTS   CURVED CAM FACE   LINK TO TERMINAL

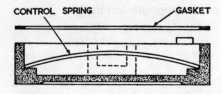

CONTROL SPRING   GASKET

Fig. 5.5   Silent alternating current switch

the switch OFF. This switch is both safe and suitable for installation in a bathroom.

All switches should be mechanically robust to withstand the constant operation, and the contacts should be heavy and firm enough to carry the current without overheating.

Plugs and socket-outlets

These are required to enable portable apparatus to be connected to the final circuits. The socket-outlet is the fixed portion connected to the fixed wiring, and comprises two or three contact tubes and terminals. The plug is the movable part connected to the apparatus by flexible wire, and comprises two or three contact pins to fit into the contact tubes. Plugs and socket-outlets are made to British Standard Specifications,

The plug pins are of phosphor-bronze or hard-drawn brass, solid or slotted down the length to form a spring contact in the tubes. The terminals are of substantial construction to clamp the flexible wire firmly. The plug cover, of tough incombustible material, is provided with a clamp to prevent the flexible wire from pulling out of the plug when in use, this clamp to be inside the cover. The

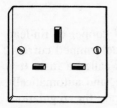

Fig. 5.6   13 A BS 1363 socket outlet

Fig. 5.7   Plug

socket-outlet base is of tough insulating material, and the contact tubes, which must be self-adjusting to the pins, are of phosphor-bronze or hard-drawn brass with sound terminals. Figure 5.6 shows a 3-pin socket-outlet for flush mounting. The earthing pin must make contact with its tube before the live terminals make contact. For direct current circuits the socket-outlet must be switch-controlled. Figure 5.7 shows a 3-pin plug made in accordance with B.S. 1363. The rating at 230 V is 3 kW, and it is suitable for use with alternating current only. The plug pins are solid, of rectangular section, and clearly identified by 'L', 'N', and 'E'. The sockets are self-adjusting. A 13-A cartridge fuse fitted inside the plug is connected between the line pin and the connection to the flexible wire. Every plug and socket-outlet shall comply with the following:

1. It must not be possible for any one pin of a plug to be in live contact with the socket-outlet while any other pin is exposed.
2. No plug should fit any live contact of a socket-outlet, other than that of the socket-outlet for which it is designed.
3. Every plug and socket-outlet must be non-reversible, with provision for earthing.
4. Every fused plug is to be non-reversible, with no provision for a fuse in an earthed conductor.

Fuses

A fuse element consists essentially of a piece of copper or tin-lead alloy wire which will melt when carrying a predetermined current. This element with contacts, carrier, and base is called a fuse. It is placed in series with the circuit to be protected, and automatically breaks the circuit when overloaded. In general, the regulations regarding fuses require that fuses shall be accessible, and shall be fitted either on the front of a switchboard or in a protecting case. In most cases in installation work the fuses are fitted in a distribution board. The position of fuses and distribution boards in an installation has already been dealt with in an earlier chapter.

Rewireable fuses

The usual type of fuse is one in which the fuse wire is carried in a removable fuse link. The fuse link may be of porcelain or other suitable insulating material, so constructed that there is no danger to the operator in removing the fuse link. The fuse wire is connected between two terminals and passes through a hole in the porcelain, or is in intimate contact with a sheet of asbestos. The fuse link is push-fitted into the fuse base to make the connection through suitable contacts. Two types of fuse are shown in Fig. 5.8.

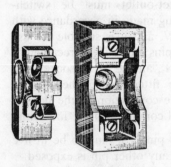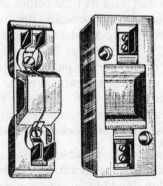

Fig. 5.8   Rewireable fuses

Although the material cost of replacing a blown fuse wire in a rewireable fuse is negligible, nevertheless this fuse has disadvantages, the chief being renewal by the wrong size of fuse wire.

Cartridge fuses

Some of the advantages of these are: quick and easy replacement, colour coding of fuse sizes, and the lack of deterioration of the fuse wire. The cartridge fuse consists of a sealed tube with metal end caps. The fuse wire passes through the tube from cap to cap and is welded or soldered to the inside of the cap. There is sometimes a blowout device in the side of the tube to indicate when the fuse is blown. When the fuse is blown the whole cartridge must be replaced. Cartridge fuses only are used in fused plugs, such as the common ring-circuit 13-A plug.

High breaking capacity fuses

This type of fuse (HBC) was brought into use some years ago to overcome the disability of ordinary fuses of destroying themselves in the event of a very heavy overload. The HBC fuse illustrated consists of a ceramic tube with metal end caps and fixing tags. The fuse is a silver strip of special shape with a low melting point rivet in the centre. The strip is entirely surrounded by chemically purified silica. When an overload occurs breaking the fuse element, the silica prevents the formation of an arc, thus preventing overheating of the fuse and its surroundings Fig. 5.9.

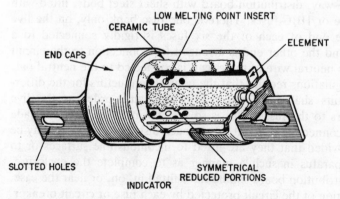

Fig. 5.9 Sectional view of HBC fuse (Dorman & Smith Ltd)

Distribution boards

By definition, the distribution board is an assemblage of parts, including one or more fuses or circuit breakers, arranged for the

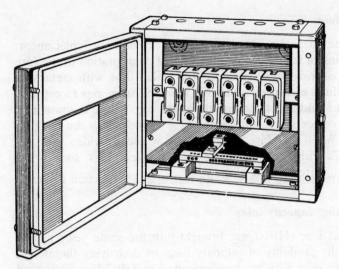

Fig. 5.10   Six-way distribution board (MEM)

distribution of electrical energy to final circuits or to other distribution boards.

The boards are usually metal-cased in sheet steel and the case is
fitted with an earthing terminal. Figure 5.10 shows a 1-pole and
neutral, 6-way, distribution board with sheet steel body, fitted with
rewireable or HBC fuses. There is one fuse bank only, on the live
side. One end of each of the six fuses is rigidly connected to a
busbar, and the other end is arranged for connection to the circuit
wire. The neutral wire of the circuit is connected to the neutral bar.

The regulations require that the neutral conductors for the differ-
rent circuits shall be connected in the same order as the live
conductors to the fuses. This is so that no mistakes shall be made
when disconnecting a circuit. Open back distribution boards may be
fitted provided that they are fitted to incombustible surfaces or to
other apparatus in such a manner as to complete the enclosure.
Every distribution board should have fitted in, on, or near the case,
an indication of the circuit protected by each fuse or circuit breaker,
and the appropriate current rating for that circuit. Where cables are
rated on the basis of close excess-current protection, it should be
indicated that the fusing factor must not exceed 1.5.

Some distribution boards are designed to contain circuit breakers
instead of fuses. These however are more expensive than the equivalent rewireable or HBC fuses.

Miniature circuit breakers

These are being increasingly used for excess-current protection in 1-phase, 250-V circuits.

The circuit breaker is essentially a switch which may be:

1. Opened and closed by hand.
2. Opened automatically when overloaded.

The tripping action may be either magnetic or thermal. In general both these actions are used in this type of circuit breaker. Protection against sustained overcurrent is given by the bending of a bi-metal strip with its time-lag effect, while high speed protection against a short circuit is given by magnetic operation.

The circuit breaker replaces both switch and fuses in the various circuits in which it is used. It can be obtained with plug-in contacts for insertion into a fuse base in a distribution board in place of a plug-in fuse carrier.

Main switch and fuses

The consumer's main switch and fuses may be combined in one case. With this type of switch and fuse gear, the switch cannot be operated when the case is open, nor can the case be opened while the switch is closed.

Consumer's control unit

In a 1-phase installation the consumer's main switch and fuses may be combined with the distribution board as one combined unit. The consumer's main fuses may be omitted if the supply authority agrees.

The unit commonly known as a Consumer's Control Unit (cca) omits the main fuses. The unit comprises a 2-pole main switch and circuit fuses. The fuses vary in size, e.g., 5 A, 15 A, and 30 A.

6 Earthing

An earth is defined as a connection to the general mass of earth. A conductor or other metal is 'earthed' when it is effectually connected to the general mass of earth by means of a metal rod or a system of metal water-pipes or other conducting object, and 'solidly earthed' when it is earthed without the intervention of a fuse, switch, cicruit breaker, resistor, reactor or solenoid.

Public electricity supply systems are invariably earthed at one point, either at the neutral point of a polyphase system, at the middle wire of a 3-wire direct current system, or on one pole of a 2-wire system. The line wires are maintained at a potential to earth, usually of 230 to 240 V.

Consider an installation wired in steel tubing, and connected to a 230-V 2-wire supply system with earthed negative. An alternative path for current from positive terminal to negative terminal is by way of an earth fault through earth to the negative terminal, as shown in Fig. 6.1.

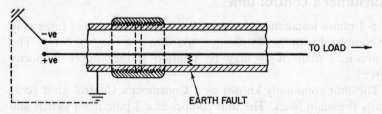

−ve

+ve

TO LOAD →

EARTH FAULT

Fig. 6.1 Illustrating earth fault

If by some mischance, perhaps by damage to the cable during drawing in, the positive wire comes into contact with the tubing, the tubing if not earthed will become charged to a potential above earth of 230 V. Any person who touches the tubing, and who is at the same time in contact with earth by standing on a non-insulating floor or who is touching earthed metal, will complete the circuit and will receive a shock the severity of which depends upon the total resistance of the circuit thus formed.

If, however, the tubing is solidly earthed, it will be at earth potential, and the heavy current that will flow through the low resistance will blow the fuses and disconnect the supply; the danger of shock is nil. A high resistance in the earth circuit, such as may be caused by a badly fitted coupling will allow part of the tubing to remain charged at a dangerous potential with consequent danger of shock to persons touching the tubing. At the same time the current may not be sufficient to blow the fuses and yet be big enough to heat the dirty connection, with consequent danger of fire.

Obviously, then, for safety, any metal liable to become charged should be earthed, and every part of the earthing circuit should be properly installed.

Bathrooms

Special precautions are to be taken in a room containing a fixed bath or shower, owing to the particularly dangerous conditions for an electric shock. As far as possible all metal parts of the electrical equipment should be shielded or hidden in the walls or ceiling. As stated earlier, on p. 20 and shown in Fig. 2.4 all metal pipes, sinks and baths (defined as extraneous conductive parts) must be bonded together by a protective conductor (supplementary conductor) which is either connected to the circuit protective conductor (c.p.c.) of a circuit supplying electrical equipment, such as a space heater, within the bathroom or is connected to the earthing terminal in the consumers control unit (c.c.u.). The object of this bonding is to reduce the value of any potential difference that may exist in the event of a fault, to a negligible value.

Lampholders should be of shrouded type and be fitted with protective shields unless totally enclosed fittings are used. Switches should be out of reach of a person in contact with the bath, or should be of the cord-operated type, or should be fitted outside the bathroom door.

No stationary appliance with heating elements which can be touched must be installed within reach of a person using the bath or shower. There must be no provision for connecting a portable appliance except a shaver supply unit complying with BS 3052. The earthing terminal of this unit must be earthed.

The residual current breaker (RCCB) or RCD is an automatically

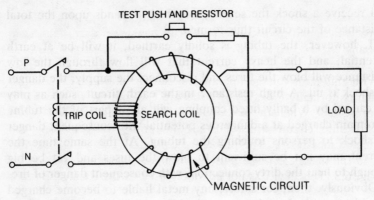

Fig. 6.2   RCD

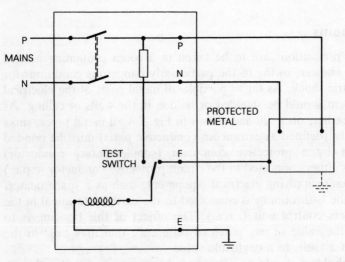

Fig. 6.3   Fault voltage operated circuit breaker

operated switch. At a predeterminal level of each earth fault current (maximum of 30 mA), the imbalance in the two primary windings will create an induced e.m.f. in the search coil (secondary) and the tripping mechanism will operate within a specified time.

Such a unit must be installed on a TT system. Also any socket outlet intended to supply portable equipment outdoors must be protected by an RCD (471–12). Figure 6.2 shows the means of connecting such an RCD.

The fault-voltage operated device is another automatic switch, which is designed to disconnect a circuit having an earth fault such that any exposed conductive parts and extraneous parts are raised to a potential of 50 V.

With this fault-voltage operated circuit breaker (FVCB), a test switch is normally in series with the trip coil and the metalwork to be protected. When the test switch is operated, it connects a test resistor into circuit whilst at the same time, disconnecting the protected metalwork. The protective conductor connecting the trip coil to the earth electrode must be insulated. Such a unit is shown in Fig. 6.3. This device is not as widely used as the RCD, the latter being commonly incorporated into consumer's units to control socket outlet circuits. It should not, as discussed earlier, be used as the main isolating device in a consumers' control unit.

7 Testing

Before any completed installation, or an addition/alteration to an existing installation, may be connected to the supply all such installation work must be inspected and tested in order to comply with Part 6 of the IEE *Wiring Regulations* (15th edition).

Any such tests that are to be made should be such that no danger to persons or damage to property can occur.

Prior to conducting any inspection and testing, the person or persons who are to undertake this task must be provided with all the drawings, charts, tables and specifications associated with the installation.

This is in order that the type and composition of circuits, protective devices used, methods of earthing, presence of sensitive electronic devices, location of switchgear and equipment may be determined.

In other words, full knowledge of the installation must be established in order to know which tests are required, and whether any special conditions apply to that particular installation. This is required by Regulation 611-2.

It is necessary then to determine that:

1. All electrical equipment installed complies with the applicable British Standard.
2. Each item of equipment, every appliance and all parts of the installation have been correctly selected (designer's responsibility) and erected/installed in order to comply with the IEE *Wiring Regulations* (15th edition).
3. All installed equipment and parts of the installation are not visibly damaged.

These three items form the objectives of the visual inspection that must be conducted before any tests are made.

The visual inspection to be made should include a check of the following items, as relevant to the installation, and is a requirement of Regulation 612-1.

1. Connection of conductors.
2. Identification of conductors.
3. Selection of conductors for current-carrying capacity and voltage drop.
4. Connection of single pole devices in phase conductor only.
5. Correct connection of socket outlets and lampholders.
6. Presence of fire barriers and protection against thermal effects.
7. Methods of protection against direct contact:
   (a) Protection by insulation of live parts.
   (b) Protection by barriers and enclosure.
   (c) Protection by obstacles.
   (d) Protection by placing out of reach.
8. Protection by non-conducting location.
9. Presence of appropriate devices for isolation and switching choice and setting of protective and monitoring devices.
10. Labelling of circuits, fuses, switches and terminals.
11. Selection of equipment and protective measures appropriate to external influences.
12. Presence of warning and danger notices.
13. Presence of diagrams, instructions and similar information.

Obviously, in order to conduct such visual inspections, it is necessary to have a logical sequence, otherwise much time may be lost by going over the ground twice or more. For the larger installation a check list would simplify matters.

Regulation 613-1 lists the sequence of the tests to be made, again appropriate to the particular installation and equipment installed.

The order and nature of tests to be made is:

1. Continuity of ring final circuit conductors.
2. Continuity of protective conductors, including main and supplementary equipotential bonding.
3. Earth electrode resistance.
4. Insulation resistance.
5. Insulation of site built assemblies.
6. Protection by electrical separation.
7. Protection by barriers or enclosures provided during erection.
8. Insulation of non-conducting floors and walls.
9. Polarity.
10. Earth fault loop impedance.
11. Operation of residual current devices and fault-voltage operated protective devices.

It is only when the visual inspections and the first eight of these tests have been carried out, that it is permissible to connect (on a temporary basis) the installation to the supply in order that the remaining three tests may be conducted.

Due to the extremely low values of resistances to be measured for continuity tests, it is important that all instruments to be used are regularly checked and re-calibrated to ensure accuracy. When recording the values for the various tests on the check list or certificate of inspection, it is advisable to record the serial number of the instrument used.

Standard methods of conduction tests are indicated in Appendix 15 of the IEE *Wiring Regulations* (15th edn), but alternative methods may be used.

Continuity of ring final circuit conductors

There are two alternative methods of applying this test. The object is to ensure that not only are the ring conductors continuous and unbroken, but are not wrongly interconnected between the outlets on the ring.

Method 1 (see Fig. 7.1)

The ends of the phase, neutral and protective conductors are disconnected from the distribution board and tested P–P, N–N, c.p.c.–c.p.c. with a low-reading ohmeter (probably a digital instrument) and the readings recorded (a).

All ends are then made together and using the long test lead, a further test is made between the ends of the joined phase, neutral and protective conductors and their respective terminations at an outlet situated midway in the ring (b) as shown in Fig. 7.1.

The resistance of the test leads is now measured (c) and this value subtracted from readings (b), resulting in a value which should be approximately one-fourth of reading (a), i.e.

$$b - c = \frac{a}{4}$$

Method 2 (see Fig. 7.2)

This method avoids the use of the extended test leads, and the test can be conducted at the distribution board.

Again a continuity test is made between the open ends of the

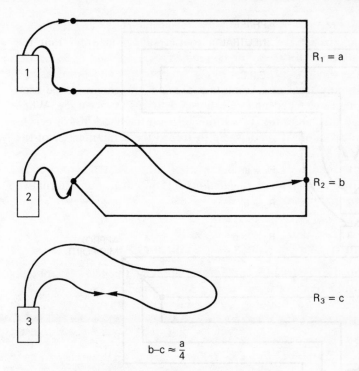

$R_1 = a$

$R_2 = b$

$R_3 = c$

$$b{-}c \approx \frac{a}{4}$$

Fig. 7.1 Method 1 test of ring final circuit continuity

phase, neutral and protective conductor at the distribution board and the values recorded.

The ends of the respective phase, neutral and protective conductor are then rejoined but not connected into the distribution board. The phase neutral and earth connections are then joined together at the socket outlet midway around the ring and the continuity test applied between the phase and neutral conductors at the distribution board.

The reading should be half of that obtained for either the phase or neutral conductor when the ring was open. Where a separate c.p.c. is used in the ring in the ring circuit, then the reading between the phase conductor and the c.p.c.'s at the distribution board should be one quarter that of the opened protective conductors.

Every protective conductor has to be tested separately to establish not only that it is electrically continuous, but also that its resistive

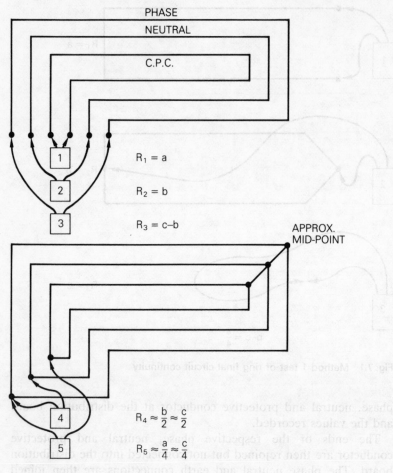

Fig. 7.2   Method 2 test of ring final circuit continuity

value (R_2) will allow for automatic disconnection in the event of an earth fault occurring.

When the protective conductor is not steel conduit or a similar ferrous enclosure, the test may be conducted with a low reading d.c. ohmeter.

However, when the c.p.c. is of steel then an a.c. test instrument should be used at mains frequency. The test voltage should not exceed 50 V and the test current should be in the order of 1.5 times the design current for the circuit under test, with a maximum of

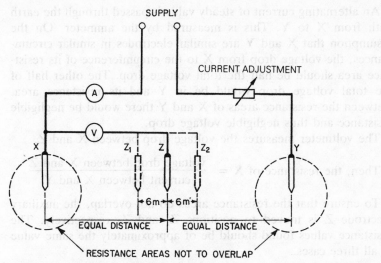

Fig. 7.3   Test of earth-electrode resistance

25 A. The test should be made from the origin of the c.p.c. (usually the distribution board) to the farthermost end of the c.p.c.

For some installations earthing has to be achieved by installing an earth electrode, and the test for the earth electronic resistance is made to ensure that the total earth loop impedance remains within the design limits.

Two methods of testing are available:

1. Using such test instruments as the 'Megger' Null-balance earth test instrument or the 'Megger' type E6 earth tester.
2. The method depicted in Appendix 15 of the IEE *Wiring Regulations* (15th edn) and as shown in Fig. 7.3.

Measurement of consumer's earth-electrode resistance

If the earth-fault loop impedance test gives too high a result, it may be necessary to measure the resistance of the earth electrode.

The resistance area of an earth electrode is the area of soil around the electrode within which a voltage gradient measurable with commercial instruments exists. In Figure 7.3 X is the earth electrode under test, and Y is an auxiliary earth electrode positioned so that the two resistance areas do not overlap. Z is a second auxiliary electrode placed halfway between X and Y.

An alternating current of steady value is passed through the earth path from X to Y. This is measured by the ammeter. On the assumption that X and Y are similar electrodes in similar circumstances, the voltage drop from X to the circumference of its resistance area should be half the total voltage drop. The other half of the total voltage drop would be in Y and its resistance area. Between the resistance areas of X and Y there would be negligible resistance and thus negligible voltage drop.

The voltmeter measures the voltage drop between X and Z.

Then, the resistance of X $= \dfrac{\text{voltage drop between X and Z}}{\text{current between X and Y}}$

To ensure that the resistance areas do not overlap, the auxiliary electrode Z is moved to positions Z_1 and Z_2 respectively. The resistance values found should be of approximately the same value in all three cases.

Insulation resistance

This is the resistance in ohms between the live parts of the installation and earth, measured *through* the insulating covering of the conductors, etc. In the case of metal-covered wiring or conduit wiring, the term 'earth' means in practice the metallic covering or conduit which itself is connected directly to earth. Additionally, the insulation resistance is measured between lines, that is, between the opposite poles of the installation with lamps or other apparatus disconnected and switches on.

The difference between insulation and conductor resistance is shown in the sketches Figs. 7.4 and 7.5. In measuring the resistance of a conductor AB, the resistance is measured along the wire from end to end, and increase of conductor length means increase of resistance. In measuring the insulation resistance of the conductor, the measurement is made from the conductor outwards.

The formula for conductor resistance is

$$R = \frac{\rho l}{a},$$

where l is the length of the conductor, a is the cross-sectional area, and ρ is the resistivity of the conductor material. Using a similar formula for insulation (it is sufficiently true for the purposes of this argument),

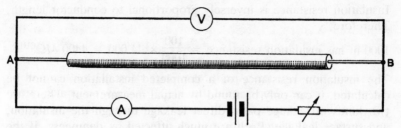

Fig. 7.4 Measuring conductor resistance

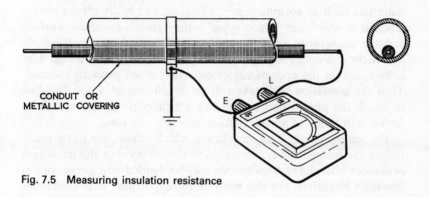

CONDUIT OR
METALLIC COVERING

Fig. 7.5 Measuring insulation resistance

$$R' = \frac{\rho' l'}{a'},$$

where l' is the *thickness* of the insulation, a' is *proportional to the length* of the wire, and ρ' is the resistivity of the insulating material.

Thus, if an insulated wire is increased in length, its conductor resistance increases while its insulation resistance decreases. Therefore, the longer the conductor the less will be its insulation resistance.

EXAMPLE

If a length of 100 metres of insulated copper cable has a conductor resistance of 1.2 Ω and an insulation resistance of 44 000 MΩ, find the corresponding values of one kilometre of the same cable.

Conductor resistance is proportional to conductor length. Therefore,

$$1 \text{ km (1000 m)} = \frac{1000}{100} \times 1.2 = 12 \ \Omega$$

Insulation resistance is inversely proportional to conductor length,
Therefore,

1000 m has insulation resistance $= \dfrac{100}{1000} \times 44\,000 = 4400 \; M\Omega$

The insulation resistance of a completed installation cannot be
calculated, it can only be found by actual measurement. There are
two kinds of leakage path: direct leakage through the insulation,
and surface leakage. Both are much affected by dampness. If the
insulating material is damp it will allow current to leak away, that
is, the insulation resistance will be low. This particularly applies to
materials such as cotton, paper, asbestos, and badly glazed porcelain, all of which are 'hygroscopic' – they absorb moisture. Surface
leakage takes place along the surface of the insulation, particularly
at switches, ceiling roses, distribution boards, etc., where dust may
collect, and at the ends of cables especially if not properly cleaned.
Thus the insulation tests taken of an installation in a new building
in which the plaster and mortar are not properly dried out will be
lower than those of an installation in a dry situation.

The various tests of an installation which follow, are to be made
before the installation is connected to the supply. For the insulation
resistance tests, large installations may be divided into groups of not
less than 50 outlets. For this purpose the expression 'outlet' includes
every point (position for attachment of lamp, lighting fitting, or
current-using appliance), and every switch and socket-outlet. A
socket-outlet, appliance or lighting fitting incorporating a switch is
regarded as one outlet.

Testing voltage

The voltage used for insulation resistance tests shall be a direct
current voltage not less than twice the normal direct current voltage,
or in the case of alternating current not less than twice the normal
r.m.s. voltage; but it need not exceed 500 V.

Thus a 500-V insulation resistance tester is sufficient for general
purposes.

Testing the installation to earth

The test shall be made with all fuse links in place, all switches
including the main switch closed and, except where earth-concentric
wiring is concerned, all poles or phases electrically connected
together.

If required, all lamps and appliances may be removed during the

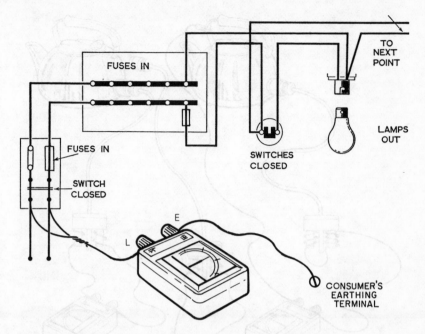

Fig. 7.6 Testing insulation resistance of completed installation

test, in which case each piece of apparatus should be separately tested. The measured insulation to earth shall be not less than 1 MΩ. The insulation resistance of each piece of apparatus measured separately shall be not less than 0.5 MΩ to earth (between live parts and frame), and 0.5 MΩ between poles or phases.

The sketch (Fig. 7.6) shows the connections for testing the insulation resistance to earth of a completed 2-wire installation with lamps and other apparatus disconnected. The wires of both poles of the supply to the main switch are twisted together and connected to the 'line' terminal of the ohmmeter. The 'earth' terminal of the ohmmeter is connected to the consumer's earthing terminal. The three terminals of 2-way switches should temporarily be connected together. Heating appliances should be tested to earth separately; (see Fig. 7.7 which shows a test on an electric kettle).

Testing between conductors

This test is made between all the conductors connected to any one pole or phase of the supply, and all conductors connected to any

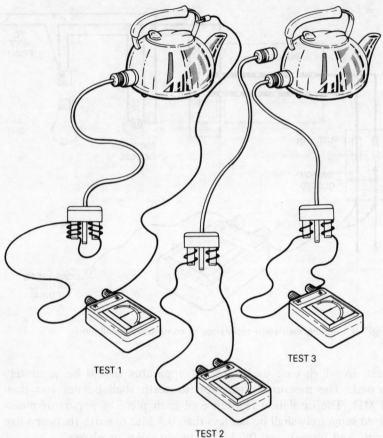

TEST 3

TEST 1

TEST 2

Fig. 7.7　Testing electric kettle

other pole or phase of the supply. The insulation resistance is to be not less than 1 MΩ.

All lamps should be removed, all current-using apparatus disconnected and all local switches controlling lamps or apparatus closed. When the removal of lamps and apparatus is not practicable, all local switches should be open. The test does not apply to earthed concentric wiring systems. Figure 7.8 gives the connections for the test of a 2-wire installation. Only one test is required in this case.

A 3-wire 3-phase installation will require three tests:

1. between lines R and Y,
2. between lines R and B,
3. between lines Y and B.

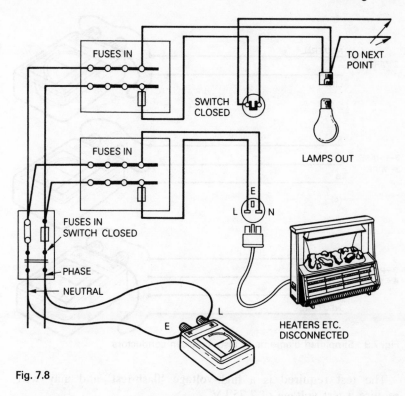

FUSES IN

SWITCH
CLOSED

TO NEXT
POINT

FUSES IN

LAMPS OUT

FUSES IN
SWITCH CLOSED

E
L   N

PHASE

NEUTRAL

HEATERS ETC.
DISCONNECTED

L

E

Fig. 7.8

A 3-phase 4-wire installation will require six tests:

1. three separate tests between pairs of lines, R-Y, R-B, and Y-B,
2. three separate tests, R-neutral, B-neutral and Y-neutral.

Figure 7.9 shows the test connections in a simplified form.

With modern-day installations, electronic devices such as dimmer switches and thyristor controls for motors are commonly used. Components within these devices are likely to be damaged or destroyed by the 500 V test voltage. It is important therefore to disconnect such sensitive devices before testing.

When insulation is applied to intended live parts within equipment that has been constructed or assembled on site, then this insulation has to meet all the tests that would have to be made by manufacturers of such equipments, in order to comply with British Standards.

This does not apply to assemblies of switchgear bought in and interconnected by insulated cables. All such manufactured equipment has been so tested.

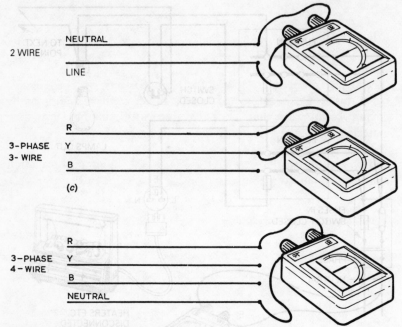

Fig. 7.9   Simplified diagrams of tests between conductors

The test required is a high-voltage 'flash-test' and may well require a test voltage of 2.75 kV.

It is most unlikely that such a test would be required for any but the most unusual of circumstances. With such 'site-built' assemblies all parts of any enclosures must be so assembled so that it is not possible for foreign objects or bodies to enter and make contact with live parts (direct contact). Therefore BS 5490 will require a degree of protection for vertical surfaces to meet IP2X and horizontal surfaces of equipment to meet IP4X.

Where special circuits are installed within a general installation, then such circuits that are fed from safety sources for example will have to be electrically separated from all other parts of that installation. Such electrical separation will have to be verified both by visual inspection and insulation and continuity tests then applied.

Barriers may well be comprised of the insulating material or sheet that prevents contact with live parts when the cover or enclosure is opened on a consumer's unit or distribution board.

These barriers and the construction of the enclosures as a whole are tested by the manufacturers. Such tests again have to meet IP2X and possibly IP4X.

For IP2X a probe (BS test finger) up to 12 mm in diameter and up to 80 mm long is used.

For IP4X a stiff straight wire of 1 mm diameter is used.

Another rare requirement that might entail measuring the insulation resistance of the floor and walls of an area would be in a special test room or a laboratory. Here persons working on live apparatus would be protected against direct contact conditions by having such walls and floor covered with suitable insulating material.

The test should be applied at three positions within such an area, and where the working voltage is 500 V the minimum insulation resistance should be 50 kΩ.

Verification of polarity of single-pole switches, etc.

It has to be ensured that all fuses and single-pole control devices are connected in the live conductor only; that the outer contacts of centre-contact bayonet and Edison-type screw lampholders are connected to the neutral or earthed conductor; and that plugs and socket-outlets have been correctly wired.

Note that I.E.E. Reg. 613-14 requires that all single-pole non-linked switches are to be fitted in the phase conductor. If the proper coloured cable is used throughout the installation, i.e. in a 2-wire installation, red for switch feeds and switch wires and black marked at both ends with a red sleeve for light feeds, no confusion should arise.

Figure 7.10 shows the means of testing with the supply disconnected.

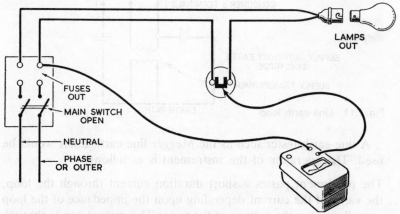

Fig. 7.10 Testing for polarity of switch with circuit dead

With supply on, an approved type test lamp is connected between the phase terminal at the switch and the circuit c.p.c. and should give full brilliance. A voltmeter similarly used will indicate full phase to earth voltage.

Earth-fault loop testing

When earth-leakage protection relies on the operation of fuses or circuit breakers, the effectiveness of earthing shall be tested by means of an earth-loop impedance test in accordance with Appendix 15.

Figure 7.11 shows the path of leakage current from an earth fault on a 2-wire final circuit. The path of the leakage current is from the earth fault (F) along the protective conductor to the consumer's earthing terminal and thence to the consumer's earth electrode. From here the fault current passes through the general mass of earth to the supply authority's earth electrode at the supply transformer, through the transformer winding and along the line through the consumer's wiring to the fault. This path is called the *line-earth loop*, and it is this loop which is to be tested.

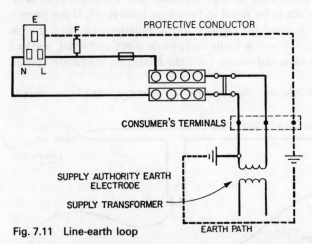

Fig. 7.11 Line-earth loop

A line-earth tester such as the Megger line-earth tester would be used. The operation of this instrument is as follows:

The instrument passes a short duration current through the loop, the value of the current depending upon the impedance of the loop as well as upon the voltage of the tester. The current passes through a 10-Ω resistor in series with the loop, and the voltage drop across it is measured by means of a ballistic instrument which is calibrated to read directly the loop impedance in ohms.

The test instrument is connected between the phase of the circuit to be tested and the circuit c.p.c. Although commonly shown connected at a socket-outlet, the test may be made at the L and E terminals of any appliance or equipment connected to a circuit. Figure 7.12 shows such a test from a socket-outlet.

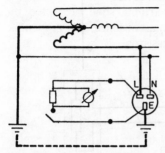

Fig. 7.12   Testing of a 3-pin socket

It is necessary to conduct this test in order to establish that the test value does not exceed the maximum value for Z_s as given in Tables 41A1 and 41A2 of the 15th edition of the IEE *Wiring Regulations*.

As residual current devices, and to a less extent fault-voltage operated devices are increasingly being installed to protect circuits and/or installations, it is necessary to test their operational efficacy. Although there is a test button incorporated in both types of device, this only checks the electro-mechanical operation of the device and not its operation under circuit/installation fault conditions.

Whilst the test button has to be operated as a periodic check, the test to be made in order to comply with the requirements is an external one, and is made by the test instrument's leads to the load connecting terminals of the RCD as shown in Fig. 7.13 (from 6.1 – Appendix 15). The test instrument has provision to allow the rated tripping current to flow. The RCD should operate within 0.2 seconds or the specified time declared by the manufacturers.

The fault-voltage operated device is tested by an instrument incorporating a double wound transformer arranged to apply up to 50 V a.c. across the neutral and frame terminal of the FVOD, which should trip instantaneously. This test is shown in Fig. 7.14 (from 6.2 – Appendix 15).

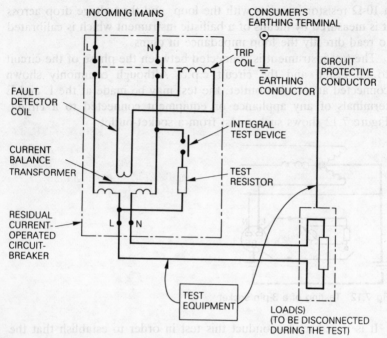

Fig. 7.13 Method of test for compliance with Regulation 613-16 for a residual current operated circuit breaker to BS 4293

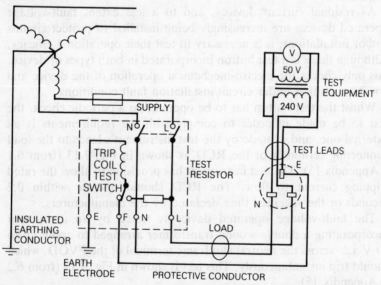

Fig. 7.14 Method of test for compliance with Regulation 613-16 for a typical fault-voltage operated protective device

Completion certificate

On completion of a new installation or of a major alteration, and after inspection and testing as described in this chapter, the installation contractor is required to give a completion certificate. This certificate, which is shown in full in the regulations, gives certain details of the installation including the number of appliances, etc., the method of earthing, and the value of the earth-loop impedance.

The certificate states that the work has been done in accordance with regulations. It also recommends periodic testing and inspection. An inspection certificate should accompany and be attached to the completion certificate. This certificate, which is much more lengthy than the previous certificate, gives the results of the full range of tests of the installation. The form of this certificate is also given in full in the regulations.

These certificates should be retained with all documents associated with the installation.

8 Instruments and measurements

Electrical quantities may be easily indicated and accurately measured by the use of suitable types of instruments, such as galvanometers, voltmeters, wattmeters, and resistance measuring devices.

Moving-coil galvanometer

This is used to indicate the passing of a current in a direct current circuit, but does not indicate definite quantities. It is used mainly in connection with laboratory testing work and for cable fault testing. The moving-coil galvanometer is in effect a small direct current motor whose armature is allowed to move over a limited range. Figure 8.1 shows the construction of the galvanometer. A soft iron cylinder C is held symmetrically between the shaped pole pieces NS of a strong permanent magnet of cobalt steel, by a non-magnetic bar B. The air gap between the pole pieces is just wide enough to allow angular movement of a rectangular coil of fine copper wire pivoted in two jewelled bearings so as to move in the gap in a clockwise direction. The current to be indicated flows from the positive terminal of the instrument through a fine hair-spring of

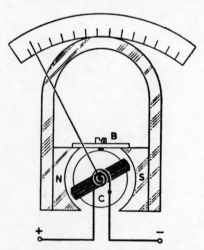

Fig. 8.1 Moving-coil galvanometer

phosphor-bronze, through the moving coil, and out to the negative terminal by a similar hair-spring. These hair-springs also control the coil movement. When the coil carries current it sets up its own magnetic field, which reacts with the magnetic field of the permanent magnet, causing the coil to move. Attached to the coil is an indicating needle which moves over a marked scale. The angular movement of the coil is proportional to the current, and the scale is equally divided. A good class instrument requires only a few milliamperes to give full-scale deflection, so the instrument must be carefully used. Since the direction of movement of the coil depends upon the direction of the current in the coil, the terminals are marked + and − to prevent incorrect connection. If the needle tends to move in a reverse direction, the circuit leads must be changed over. The moving-coil instrument may only be used with direct current.

Moving-coil voltmeter

This instrument consists of a moving coil element as described above, with a resistance of high ohmic value connected in series with it. The value of the series resistance, which is made of an alloy of negligible temperature coefficient (e.g., manganin), is such as to limit the working current to the few milliamperes necessary for full-scale deflection when the instrument is connected to a potential difference of the rated value, e.g. 250 V. The example which follows gives some idea of the value of such series resistances.

EXAMPLE

A moving-coil element of resistance 5 Ω gives full-scale deflection with a potential difference of 75 mV (0.075 V). Calculate the value of the resistance to be placed in series with the element, to give full-scale deflection when connected to a potential difference of 150 V direct current.

Current for full-scale deflection, $I = \dfrac{V}{R} = \dfrac{0.075}{5} = 0.015$ A

For same current at 150 V,

total resistance of instrument, $R_t = \dfrac{V}{I} = \dfrac{150}{0.015} = 10\ 000$ Ω

Therefore, series resistance = 10 000 − 5 = 9995 Ω

A voltmeter is always connected in parallel with the circuit whose potential difference is to be measured (see Fig. 8.2).

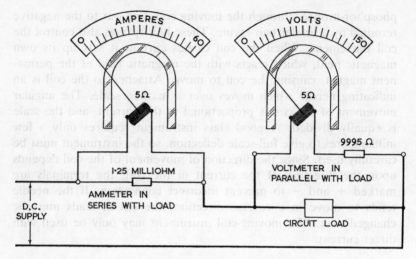

Fig. 8.2 Positions of instruments in simple circuit

Moving-coil ammeter

The moving-coil ammeter consists of a moving-coil galvanometer element connected in parallel with a shunt resistance of manganin, or similar alloy. An ammeter is connected in series with the circuit whose current is to be measured, and carries the full current. Since the current in a moving coil is very small, the shunt resistance is designed to take the whole current less the element current. The example which follows gives an idea of the ohmic value of such shunt resistances.

EXAMPLE

The moving-coil element of the last example, of resistance 5 Ω, gives full-scale deflection with a potential difference of 75 mV. Calculate the value of the shunt resistance to enable the instrument to read up to 60 A.

Total volts drop across terminals = 0.075 V
total current = 60 A
Therefore,

total resistance $= R_t = \dfrac{V}{I} = \dfrac{0.075}{60} = 0.001\,25\ \Omega$

but $\dfrac{1}{R_t} = \dfrac{1}{5} + \dfrac{1}{x}$

Therefore, $\dfrac{1}{x} = \dfrac{1}{R_t} - \dfrac{1}{5} = \dfrac{1}{0.001\,25} - \dfrac{1}{5} = \dfrac{4000 - 1}{5} = \dfrac{3999}{5}$

Therefore, $\qquad x = \dfrac{5}{3999} = 1.2503\ m\Omega$

$$= \dfrac{1.2503}{1000}\Omega$$

Moving-iron instrument

There are two types of moving-iron instrument:

1. the repulsion type, and
2. the attraction type.

The repulsion type of M.I. instrument (Fig. 8.3)

A coil of wire carried on a cylindrical non-magnetic former carries the current in the instrument. A curved 'iron' of soft iron or of mumetal is fixed to the inside of the bobbin former and is magnetized when the coil carries current. Another curved 'iron' is mounted

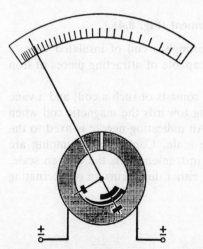

Fig. 8.3 Moving-iron instrument; repulsion type

on a spindle which passes axially through the coil, and at rest is very close to the fixed iron. This iron is also magnetized when current flows in the coil, and is of the same polarity as the fixed iron. Repulsion takes place between the two irons, causing the moving iron to make an angular movement about the spindle in a clockwise direction. An indicating needle is fixed to the spindle so that the movement is registered on a scale. Control of the movement is either by hair-springs, when the instrument can be used in any position, or by balance weights, when the instrument must be operated vertically. A 'damping' device is necessary to prevent swinging of the needle, and this consists usually of an aluminium vane fitting very closely in an air chamber. The repulsion of the irons is proportional to the square of the current, and thus the scale is uneven, being crowded at the lower values and open at the higher values. The irons may be so shaped and arranged that the scale is opened out to some degree at the lower values of current. If the direction of current in the magnetizing coil is reversed, the polarity of both irons is reversed and repulsion takes place in the same direction as before. Thus the instrument may be used either with direct current or with alternating current. The moving-iron instrument is both cheaper and more robust than the moving-coil instrument, but it has not the same degree of accuracy. The power taken by the instrument is also greater than that taken by the corresponding moving-coil instrument.

The attraction type of M.I. instrument (Fig. 8.4)

When direct current is passed through a coil of insulated wire, a magnetic field is set up which is capable of attracting pieces of iron or steel into the coil.

The attraction type instrument consists of such a coil, and a vane of soft iron so pivoted as to swing towards the magnetic coil when current passes through the coil. An indicating needle is fixed to the vane and registers on a suitable scale. Control and damping are similar to those of the repulsion instrument as is the uneven scale. The instrument may be used with either direct current or alternating current.

Moving-iron ammeter

The whole current passes through the ammeter winding, and a shunt is not required except with multi-range ammeters. For heavy

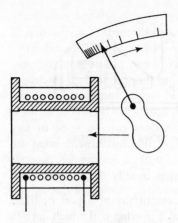

Fig. 8.4   Moving-iron instrument; attraction type

currents, the magnetizing coil consists of a few turns only of heavy section wire, while for low currents the coil consists of many turns of small section wire.

Moving-iron voltmeter

The lower range voltmeters can be wound so that series resistances are not needed. In the higher voltage ranges, series resistances within the voltmeter case are necessary.

The wattmeter

This is an instrument for measuring the power in a circuit. Since in a direct current circuit the power is the simple product of volts and amperes, $P = VI$, a wattmeter is unnecessary and is seldom used. With alternating current, however, the power is not the simple product of volts and amperes (except at unity power factor), and is given by the formula, $P = VI \cos \phi$. The turning effort of a watt-meter element at any instant is proportional to the product of current and voltage at that instant, and the average turning effort is a measure of the average value of the power (see Fig. 8.5). There are two main types of wattmeter: the dynamometer wattmeter used for both direct current and alternating current, and the induction type wattmeter used only on alternating current.

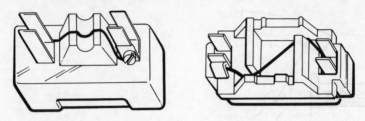

Fig. 8.5   Diagram showing power in the alternating current circuit

The dynamometer wattmeter

The dynamometer wattmeter consists essentially of a pair of fixed coils which carry the main current, and a moving coil which carries a small current proportional to, and in phase with, the voltage of the circuit (see Fig. 8.6). The fixed and moving coils are connected in the circuit, as shown in Fig. 8.6. The magnetic fields of the fixed and moving coils react on one another, causing the moving coil to turn about its axis. The movement is controlled by hair-springs which also lead the current into and out of the moving element. Damping is by light aluminium vanes moving in an air dashpot. The

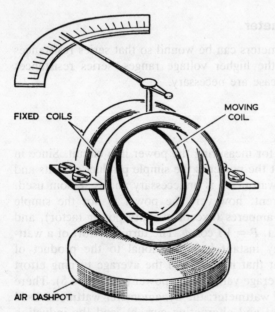

FIXED COILS

MOVING COIL

AIR DASHPOT

Fig. 8.6   Dynamometer wattmeter

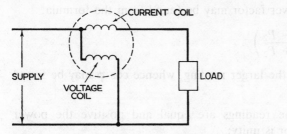

Fig. 8.7 Theoretical diagram of wattmeter

indicator needle is fixed to the moving-coil spindle, and moves over a suitably calibrated scale. For single-phase working one element only is required. For a 3-phase balanced circuit also, one element only is necessary, with a suitably calibrated scale. For 3-phase unbalanced circuits two elements are required, in accordance with the 'two-wattmeter' principle. The elements are insulated from each other and mounted one above the other. A central spindle carries both moving coils. The total torque or turning effort on the spindle is the algebraic sum of the separate torques.

Two-wattmeter method of measuring power

The power in a 3-phase balanced circuit may be measured by two wattmeters connected as shown in Fig. 8.8. Reference to textbooks on electrotechnology will show that the total power is the sum of the two wattmeter readings. Assuming that the two readings are P_1 and P_2, the total power is

$P = P_1 + P_2$

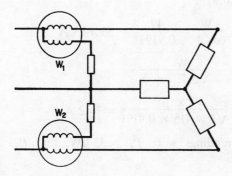

Fig. 8.8

In addition the power factor may be found from the formula,

$$\tan \phi = \sqrt{3} \left(\frac{P_1 - P_2}{P_1 + P_2} \right)$$

assuming P_1 to be the larger reading, whence $\cos \phi$ may be found from the tables.

Furthermore: if the readings are equal and positive the power factor is unity;

at a power factor of 0.866, one reading is double the other;

at a power factor of 0.5, one reading is zero;

and at a power factor below 0.5, one reading is negative.

EXAMPLE
The steady readings of two single-phase wattmeters in a 3-phase circuit are 35 kW and 14 kW respectively.
   Calculate:

   (i) the power taken by the load,
   (ii) the power factor,
   (iii) the load in kVA, and
   (iv) the line current, if the supply is at 415 V.

   (i) Power in kW $= 35 + 14 = 49 \text{ kW}$

   (ii) Power factor. $\tan \phi = \sqrt{3} \left(\dfrac{35 - 14}{35 + 14} \right) = \dfrac{21}{49}$

$$= 0.4285$$

   From tables, $\cos \phi \;\; = 0.9191$

   (iii) Load in kVA $= \dfrac{\text{kW}}{\cos \phi} = \dfrac{49}{0.9191} = 53.4 \text{ kVA}$

   (iv) Line current $I = \dfrac{\text{W}}{\sqrt{3} \text{ E} \cos \phi}$

$$= \dfrac{49 \times 10^3}{\sqrt{3} \times 415 \times 0.9191} = 74.2 \text{ A}$$

If in the above case, the readings were $P_1 = 35$ kW, and $P_2 = -14$ kW, then

   (i) Power in kW $= 35 - 14 = 21 \text{kW}$

(ii) Tan φ $= \sqrt{3}\left(\dfrac{35 - (-14)}{35 + (-14)}\right) = \dfrac{49}{21} = 2.333$

From tables, cos φ = 0.4083

(iii) Load in kVA $= \dfrac{21}{0.4083} = 51.4$ kVA

(iv) Line current $= \dfrac{21 \times 10^3}{\sqrt{3} \times 415 \times 0.4083} = 71.6$ A

The induction type wattmeter

The induction type wattmeter consists of a voltage coil and current coils. The voltage coil is wound on a laminated iron core forming a nearly closed magnetic circuit. The circuit is thus highly inductive, and the current in the coil lags about 85° behind the voltage. The current coils are wound on an open magnetic circuit which has very little inductance. A light aluminium disc is mounted in front of the magnets so that its rim will pass between the two pairs of poles. Each pair of poles sets up eddy currents in the disc. These eddy currents react on each other, causing the disc to move. The control is by hair-springs. Damping is provided by means of a permanent magnet which sets up retarding eddy currents in the disc during movement. Figure 8.9 is a diagram of a single-element wattmeter.

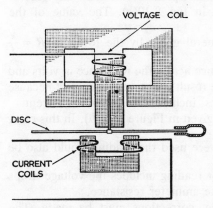

Fig. 8.9 Induction type wattmeter

Energy meters or supply meters

Such meters are necessary in order to register the energy supplied to an electricity consumer over a given period. Energy is the product of power and time, and a wattmeter element whose movement is allowed to continue and move a train of gears with indicating fingers constitutes an energy meter when properly calibrated. There are various types of energy meter:

1. Mercury motor ampere-hour meters, direct current, calibrated to read kilowatt-hours at the declared voltage.
2. Electrolytic ampere-hour meters, direct current, or rectified alternating current, calibrated as type 1, to read kilowatt-hours or kilovolt-ampere-hours.
3. Induction type energy meters, alternating current.

Measurement of resistance

Methods of measuring resistance are:
1. by ammeter and voltmeter,
2. by Wheatstone Bridge, and
3. by the direct-reading ohmmeter.

Ammeter and voltmeter method

Ohm's law may be applied to a complete circuit or to any part of a circuit. Consider the circuit in Fig. 8.10(a). The value of the resistance R in ohms is in general given by the formula, $R = \dfrac{V}{I}$ where V is the potential difference across the resistance in volts and I is the current in amperes. The result is not exactly correct because the ammeter reading, I amperes, includes the voltmeter current.

An alternative connection is given in Figure 8.10(b). In this case, if the simple formula $R = \dfrac{V}{I}$ were used the reading would also be incorrect, because the voltmeter reading includes the voltage drops across both resistance R and the ammeter resistance.

In both cases, for exactitude, corrections must be made. The example which follows shows this.

EXAMPLE

In both circuits Fig. 8.10 (a) and Fig. 8.10 (b), the resistances of the ammeter and of the voltmeter are known to be 0.05 Ω and 35 Ω

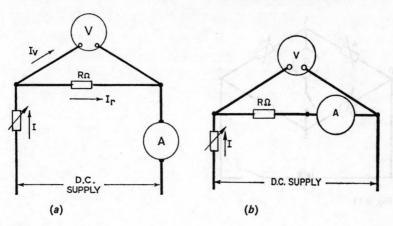

Fig. 8.10

respectively. In both cases the ammeter and voltmeter readings are 5 A and 35 V respectively. Calculate the resistance in Ω.

The normal calculation in both cases is $R = \dfrac{V}{I}$

Therefore, $R = \dfrac{35}{5} = 7 \; \Omega$

Correction in case (a):

Voltmeter current $I_v = \dfrac{35}{350} = 0.1$ A

Therefore, current in resistance $= 5 - 0.1 = 4.9$ A

Then, true resistance $R = \dfrac{35}{4.9} = 7.14 \; \Omega$

Correction in case (b):

Voltage drop across ammeter $v_a = 0.05 \times 5 = 0.25$ V

Therefore, voltage drop across resistance $R = 35 - 0.25 = 34.75$ V

Then, true resistance $R = \dfrac{34.75}{5} = 6.95 \; \Omega$

Careful consideration of the above will show that when using the uncorrected formula $R = \dfrac{V}{I}$, the results will be more nearly accurate for low values of resistance when connection (a) is used, and for high values of resistance when connection (b) is used.

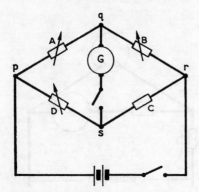

Fig. 8.11

Wheatstone Bridge

Suppose four resistances A, B, C, and D are connected as in Fig. 8.11 with a battery maintaining a potential difference across p and r, and with a galvanometer connected between q and s. If the galvanometer switch is open, the current flowing from the battery divides, part passing through A and B in series, and the remainder through D and C in series. The potentials of the points q and s are lower than the potential of p, and higher than the potential of r. Since the potential drops in two resistances in series are proportional to the resistances, then,

$$\frac{\text{potential drop across A}}{\text{potential drop across B}} = \frac{\text{resistance of A}}{\text{resistance of B}}$$

Also, $\dfrac{\text{potential drop across D}}{\text{potential drop across C}} = \dfrac{\text{resistance of D}}{\text{resistance of C}}$

If q and s are at the same potential, no current will flow in the galvanometer when the galvanometer switch is closed. Assuming this to be the case, then,

$$\frac{\text{potential drop across A}}{\text{potential drop across B}} = \frac{\text{potential drop across D}}{\text{potential drop across C}}$$

Therefore, in terms of resistances, $\dfrac{A}{B} = \dfrac{D}{C}$

If we know any three of these quantities the fourth may be calculated:

thus, if A, B, and D are known, $C = \dfrac{BD}{A}$.

The arrangement is known as the Wheatstone Bridge, and is used to find resistance values to a high degree of accuracy. The resistances A, B and D are separately adjustable, and C is the resistance whose value is to be found. The three adjustable resistances are varied until no current flows in the galvanometer. The bridge is then said to be 'balanced', and

the formula given above is applied, i.e. $C = \dfrac{BD}{A}$.

The *Slide Wire* form of Wheatstone Bridge is shown in Fig. 8.12. A uniform resistance wire 1 metre long is stretched between two busbars p and r. The galvanometer G is connected between a busbar q and a slider connection s. A known resistance D and the unknown resistance C are connected as shown. The connection s is moved along the wire until the bridge is balanced. The resistances of the wire sections A and B are proportional to their lengths, and therefore these lengths may be used in the formula.

Then $\qquad \dfrac{\text{resistance D}}{\text{resistance C}} = \dfrac{\text{length A}}{\text{length B}}$

Therefore, resistance $C = \dfrac{D \times \text{length B}}{\text{length A}}$

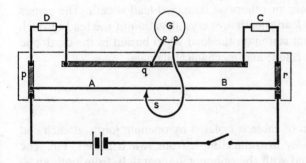

Fig. 8.12   Slide wire bridge

EXAMPLE

Resistance D = 7 Ω, and at balance A = 300 mm and B = 700 mm.

Then $\qquad\qquad C = 7 \times \dfrac{700}{300} = \dfrac{49}{3} = 16.3 \ \Omega$

Testing for faults on lead-sheathed cables

The tests described below are commonly carried out by mains engineers on faulty feeder cables and distributors, and may be applied with equal success by the installation engineer or the works electrician. The tests are obviously only required on the long lengths of cable where position of the fault is not visible to the eye.

Types of fault

(a) **Earth fault**. This is most common; the conductor is in contact with the lead sheath, and the fault resistance may be low or high.

(b) **Short circuit between cores**. This is much less common than the earth fault, and is usually found in combination with an earth.

(c) **Open circuit, or broken conductor**. A clear break is seldom met with, and only occurs when the cable has been unduly stretched by accident.

Causes of faults

Most faults are caused by dampness in the paper insulation of the cable due to porous or otherwise damaged lead sheath. The causes of damage to the lead sheath are: crystallization of the lead through vibration; chemical action on the lead when buried in the earth and insufficiently protected; and mechanical damage.

Procedure

The faulty section of cable is isolated by opening joints at each end of the length, or by disconnecting it from switch panels, etc. The cable is then tested with the ohmmeter separately from both ends. The tests made are of each core to earth, and between each pair of cores. Continuity tests are also necessary. Making a diagram illustrating the results will greatly assist a correct diagnosis.

EXAMPLE

Show by diagram the fault condition of a 3-core cable whose insulation resistance measurements are given below:

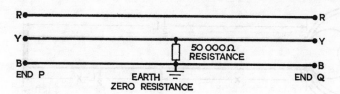

Fig. 8.13

From end P –
Red core to earth: infinity   Red to yellow: infinity
Yellow core to earth: 50 000 Ω   Red to blue: infinity
Blue core to earth: zero   Yellow to blue: 50 000 Ω

From end Q –
Similar results.

Continuity tests show that the cores are fully continuous.

The diagram shows the fault condition (Fig. 8.13). The red core is in good condition throughout, there is a 50 000-Ω fault between the yellow and blue cores, and there is a zero resistance earth on the blue core.

Fault-location tests

Murray loop tests ⎫
Fall of potential tests ⎬ for each faults and short circuits
Capacity tests for open circuits.

The Murray loop tests

This is a variation of the slide wire bridge already dealt with. Assume that one core of a twin cable has a low resistance fault to earth, and that the other core is in good electrical condition. Figure 8.14 shows the connections. A loop of resistance wire of known length 100 units long (centimetres) and a galvanometer or a milli-ammeter are connected in parallel to the near ends of the two cores. The connections should be side by side and not one on top of the other. The two cores are short circuited at the far end by a heavy shorting strip of negligible resistance. For preference the cores should be sweated together. A battery is connected by slider

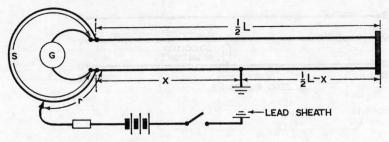

Fig. 8.14 Connections for Murray loop test

connection to the loop. If the fault is of very low resistance, 6 V are sufficient. The other end of the battery circuit is connected to earth, i.e. to the lead sheath of the cable, with the positive of the battery to the fault. The test is carried out by moving the sliding connection along the loop until the balance point is found and the galvanometer reading is zero.

Let the *length of the complete loop* of cable be L metres, the distance from the near end to the fault be x metres, the whole length of the slider loop be s units, and the distance of the slider from the end of the faulty core be r units.

Then

$$\frac{r \text{ units}}{s \text{ units}} = \frac{x \text{ m}}{L \text{ m}}$$

Therefore,

$$\frac{r}{s} = \frac{x}{L}$$

and

$$x = \frac{rL}{s} \text{ m}$$

As a check, the test should be repeated from the far end of the cable.

If the second core of the cable is also faulty, the test can be made as shown in Fig. 8.15, the cable loop being made with any available cable. If the section of the cable is different from the faulty cable, the 'equivalent length' of the cable must be calculated and used in the formula.

EXAMPLE
An earth fault occurs on one core of a 400-m length of 35 mm^2 twin cable. A loop test is made. The slide wire is 1 metre long, and the balance point is 300 mm from the end connected to the faulty core.

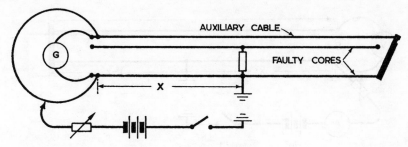

Fig. 8.15   Murray loop test, using auxiliary cable

Find the position of the fault. Refer to Fig. 8.14.
Length of cable loop $L = 2 \times 400 = 800$ m

Then $\qquad \dfrac{r}{s} = \dfrac{x}{L}$

Therefore, $\qquad \dfrac{300}{1000} = \dfrac{x}{800}$

Therefore, $\qquad x = \dfrac{300 \times 800}{1000} = 240$ m from the testing end

EXAMPLE
An earth fault occurs on a 400-m length of 25 mm^2 twin cable, both
cores being earthed at the same point. A loop is made, the cable
loop being made by means of a 500-m length of 50 mm^2 cable. The
slide wire is 1 metre long, and the balance position is 550 mm from
the end attached to the faulty core. Find the position of the fault.
Refer to Fig. 8.15.
Equivalent length of return cable (increase of section means
decrease of resistance)

$$= 500 \times \frac{25}{50} \text{ m} = 250 \text{ m}$$

Therefore, TOTAL length of cable loop $= 400 + 250 = 650$ m

Therefore, $x = \dfrac{rL}{s} = \dfrac{550}{1000} \times 650 = 357.5$ m from the testing end

Fall of potential test

Assume a twin cable of length L metres, with an earth fault x metres
from the near end P. The test circuit is shown in Fig. 8.16, with a

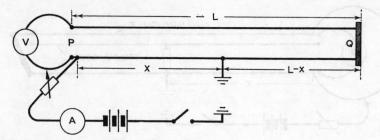

Fig. 8.16   Fall of potential test

heavy short circuiting strip connected to the two cores at end Q. The battery circuit serves to pass current through the length of cable between P and the fault. This current is registered by the ammeter. The voltmeter measures the potential drop across the ends of this part of the cable. The current through the fault is adjusted by means of an adjustable resistance to suit the range of the instruments. The resistance of the fault itself plays no part in the test, except to tend to limit the test current. Readings of volts and amperes are taken. Let these readings be V_1 and I_1. The apparatus is then moved to the end Q, and the whole test repeated with the end P now short circuited. Let these readings be V_2 and I_2.

Then *if L metres is the length of the faulty core,*

$$x = \frac{LV_1I_2}{V_2I_1 + V_1I_2} \text{ metres}$$

Note that the resistance of the return wire is not included in the formula, and thus wire of any suitable size and section may be used for this return.

Proof: Potential drop, $V_1 = RI_1 = \dfrac{\rho x I_1}{a}$

Therefore, $x = \dfrac{aV_1}{\rho I_1}$

Also, potential drop, $V_2 = \dfrac{\rho(L - x)I_2}{a}$

Therefore, $L - x = \dfrac{xaV_2}{\rho I_2}$

Therefore, $L - x = \dfrac{aV_1}{\rho I_1} \div \dfrac{aV_2}{\rho I_2} = \dfrac{V_1 I_2}{V_2 I_1}$

Therefore, $x = \dfrac{(L - x)V_1 I_2}{V_2 I_1} = \dfrac{LV_1 I_2 - xV_1 I_2}{V_2 I_1}$

Therefore, $xV_2 I_1 = LV_1 I_2 - xV_1 I_2$

Therefore, $x(V_2 I_1 + V_1 I_2) = LV_1 I_2$

Therefore, $x = \dfrac{LV_1 I_2}{V_2 I_1 + V_1 I_2}$

EXAMPLE

A twin cable of length 800 metres has developed a low resistance earth on one core. A fall of potential test is taken from both ends, P and Q. The readings taken at P are $V_1 = 1.5$ V, and $I_1 = 5$ A. The readings at Q are $V_2 = 3$ V, and $I_2 = 6$ A. Find the position of the fault.

Distance from end P in m:

$$x = \frac{LV_1 I_2}{V_2 I_1 + V_1 I_2} = \frac{800 \times (1.5 \times 6)}{(3 \times 5) + (1.5 \times 6)}$$

$$= \frac{800 \times 9}{15 + 9}$$

Short-circuit fault

Both the loop test and the fall of potential test may be used for a short-circuit fault. Figure 8.17 shows the connections for a loop test on a 3-core cable, which has a short circuit without earth between

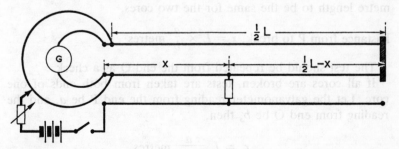

Fig. 8.17

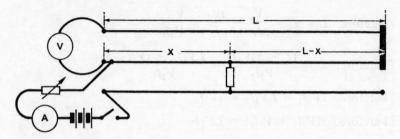

Fig. 8.18

two cores, the third core being undamaged. The battery circuit is connected between the slider connection and one of the faulty cores. Apart from this connection the whole test is as for the earth test, and the same formula is used.

The fall of potential test could also be used with the necessary alteration of connections (Fig. 8.18).

Open-circuit fault

The testing method used is to compare the electrostatic capacitance of the cable core up to the break, with the capacitance of an undamaged core. Figure 8.19 shows the connections for testing a twin cable of length L metres of which one core PQ is broken. The other core RS is sound. A charge and discharge key is connected to P, and the cable is charged for a definite period of time, say, 10 seconds, by making contact to the battery. The cable is then discharged through a ballistic galvanometer, and the deflection a is noted. The sound core RS is then charged for the same period of time and discharged through the ballistic galvanometer, and the deflection b noted. Then, assuming the capacitance to *earth* per metre length to be the same for the two cores,

Distance from P to break, $x = L \times \dfrac{a}{b}$ metres

The test should be repeated from the end Q as a check.

If all cores are broken, tests are taken from both ends of one core. Let the galvanometer reading from the end P be a, and the reading from end Q be b, then,

$$x = L \frac{a}{a + b} \text{metres}$$

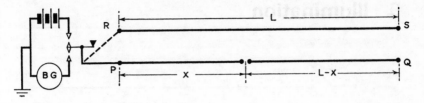

Fig. 8.19 Test for open-circuit fault

The essential requirement for these capacitance tests is that the core to be tested has a good insulation resistance.

The Bridge Megger testing set

This type of instrument is very useful to maintenance and test engineers. It is a hand-driven instrument which combines the functions of the insulation tester and the Wheatstone Bridge.

It is a true ohmmeter, enabling the insulation resistances of circuits and machines to be tested, as in the previous chapter. The incorporated Wheatstone Bridge allows the resistance values of motor windings, contactor coils, starting rheostats and the like, to be quickly and precisely determined, either as routine tests or as breakdown tests.

Faults in cables can be tested by means of the Murray Loop and other similar tests. Basically the instrument generator replaces the battery normally needed for these particular tests. Various connections are possible, in some of which the larger part of the test circuit is arranged external to the testing instrument, and others in which the whole of the test circuit, apart from the faulty cable itself, is contained inside the set.

9 Illumination

Light is a form of radiant energy. It may be produced by electric currents passed through filaments as in the incandescent lamp, through arcs between carbon or metal rods, or through suitable gases as in neon and other gas tubes. In some forms of lamps, the light is due to fluorescence excited by radiation arising from the passage of electricity through mercury vapour.

Most bodies reflect light in some measure, and when illuminated from an original source they become secondary sources of light. A good example is the moon, which illuminates the earth by means of reflected light originating in the sun.

Illumination by reflected light is of great practical importance; electric lamps are rarely used without reflectors, and light reflected from the walls and ceiling of a room makes an important contribution to the illumination of the room.

Law of inverse squares

If a source of light which sends its light out equally in all directions be placed at the centre of a hollow sphere, the light will fall uniformly on the inner surface of the sphere; that is to say, each square metre of the surface will receive the same amount of light. If the sphere be replaced by one of larger radius, the same total amount of light is spread over a larger area proportional to the square of the radius. The amount which falls upon any square metre of such a surface will therefore diminish as the radius increases, and will be inversely proportional to the square of the radius.

A similar relation holds if we have to deal with a beam of light in the form of a cone or a pyramid, as in Figure 9.1. If we consider parallel surfaces which cut the pyramid at different distances from the source, the areas of these surfaces are proportional to the squares of these distances, and therefore the amount of light which falls on one unit of the area of these surfaces is inversely proportional to the square of the distance from the source. This relation-

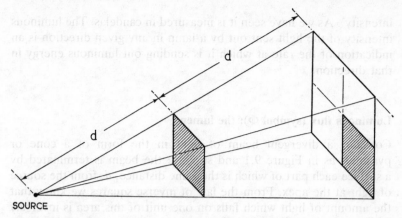

SOURCE

Fig. 9.1 Illustrating the Law of inverse squares

ship is referred to as the LAW OF INVERSE SQUARES.

When light falls on a surface like that of a piece of white blotting-paper or a piece of plaster of Paris, the brightness of the surface as it appears to the eye, viewing it from a fixed distance, is proportional to the amount of light which falls upon each unit of area of the surface. To produce the same brightness on a given surface, a strong light would have to be placed farther away from it than a weaker one. This fact, together with the law of inverse squares, enables us to construct an instrument called a photometric bench for comparing the strengths of different sources of light. This instrument is described in a later paragraph.

It is not sufficient to be able to say that the strength of one source of light is so many times greater than that of another; we need to be able to measure the strengths in agreed units. The agreed unit in this country is the candela, and the standards are kept at the National Physical Laboratory. Standard and sub-standard lamps of many sizes and types can be purchased from lamp manufacturers. They can be used as standards of comparison and for calibrating photometric benches.

Terms used in illumination

Luminous intensity (symbol I)

In the preceding paragraph we have spoken of 'the strength of a source of light'. The technical term for this quantity is 'luminous intensity'. As we have seen it is measured in candelas. The luminous intensity of the light sent out by a lamp in any given direction is an indication of the rate at which it is sending out luminous energy in that direction.

Luminous flux (symbol Φ): the lumen

Consider a divergent beam of light in the form of a cone or pyramid, as in Figure 9.1 and suppose the beam is terminated by a surface each part of which is the same distance, d, from the source of light at the apex. From the law of inverse squares we know that the amount of light which falls on one unit of this area is inversely proportional to d^2, and if the whole area is A and the luminous intensity of the source is I, the amount of light falling on A will be proportional to $\frac{AI}{d^2}$. The technical term for 'amount of light' is 'luminous flux', and the unit is the 'lumen'.

If A and D are measured in the same system of units, that is, A in square metres and d in metres, then the amount of light falling on A will be $\frac{AI}{d^2}$ lumens. If A, I, and d were all unity, the amount of light would be 1 lumen, i.e. a lumen is the luminous flux falling on unit area illuminated by a source with a luminous intensity of 1 candela unit distance away.

If the beam of light were a narrow one with a small angle at the apex, the ratio $\frac{A}{d^2}$ would be small, but if it were a very divergent one $\frac{A}{d^2}$ would be large. This ratio is called the solid angle included by the beam, and it is unity if $A = d^2$. A lumen is therefore the luminous flux put into a unit solid angle by a light source of 1 candela.

Illumination (symbol E)

When we throw light upon an object, the effectiveness of the illumination will depend on the number of lumens the object receives and particularly upon the number of lumens per unit of its surface. If the object receives I lumens on an area A, the illumination is said to be $E = \frac{I}{A}$. If A is in square metres then E is in lumen/m^2, or in lux. The lux is the unit of illumination.

We have just seen that an area A illuminated by a source of I candelas at a distance d receives $\dfrac{I}{A}$ lumens. Each unit of that area therefore receives d^2 lumens, and if d is in metres the illumination is $E = d^2$ lux.

Cosine law

Very often the illuminated surface is not normal to the direction of the light as AC in Fig. 9.2 but is inclined as AB. The area over which the light is spread is then increased in the ratio

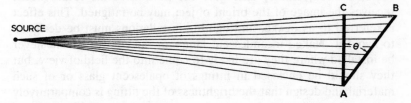

Fig. 9.2 Illustrating the Cosine law

$$\frac{AB}{AC} = \frac{1}{\cos \theta}.$$

and the illumination is decreased in the ratio $\dfrac{\cos \theta}{1}$. The expression for the illumination then becomes

$$E = \frac{I \cos \theta}{d^2}$$

Brightness

When the eye receives a great deal of light from an object we say it is bright, and 'brightness' is an important quantity in illumination. It is all the same whether the light is produced by the object or merely reflected from it. The object sends out light as though each small piece of its surface were of a certain luminous intensity. Generally, the brightness of an object is not the same from all points of view.

When brightness is considered in terms of measurable quantities, the term 'luminance' is employed. The luminance of a surface, which may be either a source or a reflecting surface is given in candelas per square metre (cd/m²).

Glare

The size of the opening of the pupil in the human eye is controlled by its iris. If the eye looks at a bright object such as a naked lamp which sends a large amount of light into the eye and produces an intense image on the retina, the iris automatically contracts to protect the eye by reducing the intensity of the image. If the eye is looking at some other object, much less bright, while the very bright object remains in the field of view, the contraction of the iris will cut down the amount of light received on the retina from every object in the field of view and make it more difficult to see the object desired; at the same time the portion of the retina which receives the image of the bright object may be fatigued. This effect is referred to as GLARE, and lighting installations must be designed to avoid it. Naked incandescent lamps or naked arcs should never be installed where they are likely to come into the field of view, but they should be enclosed in fittings of opalescent glass or of such material and design that the brightness of the fitting is comparatively low.

Photometry

The common laboratory apparatus used in comparing the luminous intensities of different lamps is the photometric bench, which consists of a horizontal frame scaled in centimetres, upon which are movable blocks which carry the photometer head, a standard lamp whose luminous intensity is known and the lamp whose luminous intensity is to be measured (Fig. 9.3). The photometer head, of which there are several kinds, consists essentially of two adjacent similar surfaces which may be observed together and upon one of which falls the light from the standard lamp, while the light from the test lamp falls upon the other surface. The photometer head is moved about along the frame until the two illuminations appear to the observer to be equal. When the photometer is thus 'balanced', the luminous intensity of the test lamp may be calculated from the formula

$$\frac{\text{luminous intensity of unknown lamp}}{\text{luminous intensity of standard lamp}} = \frac{BC^2}{AB^2}$$

or

luminous intensity of unknown lamp = luminous intensity of standard lamp $\times \dfrac{BC^2}{AB^2}$

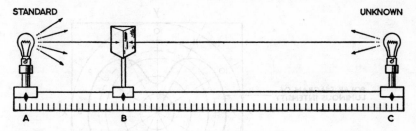

Fig. 9.3 Photometric bench

EXAMPLE

Suppose the luminous intensity of the standard lamp is 100 candelas, the distance AB is 120 cm and the distance BC is 180 cm at balance. Find the luminous intensity of the lamp on test.

$$\text{Unknown luminous intensity} = 100 \times \frac{180^2}{120^2} = 225 \text{ cd}$$

Methods of denoting luminous intensity

The illumination from any one type of lamp is not uniform in all directions, as a point source of light is unobtainable in the practical lamp. Three methods of denoting luminous intensity may be noted.

Mean horizontal luminous intensity

In Fig. 19.4 the lamp on test is shown in a vertical position, lamp cap downwards. The lamp may be rotated about its vertical axis in

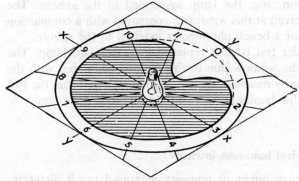

Fig. 9.4 Illustrating mean horizontal luminous intensity

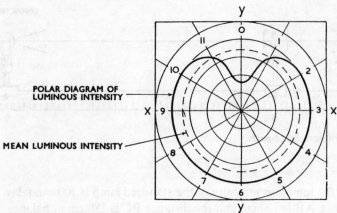

Fig. 9.5   Polar diagram

a series of equal steps and corresponding readings taken. The average luminous intensity obtained from these readings during one revolution is called the mean horizontal luminous intensity (Figs 9.4 and 9.5).

Mean spherical luminous intensity

This is the average luminous intensity measured in all directions from the source. It is not usually measured by means of the bench photometer directly owing to the large number of readings which would be required. An integrating sphere photometer is used, consisting of a hollow sphere with a matt white interior surface. The lamp to be tested is suspended inside the sphere. A small milk-glass window in the surface of the sphere is shielded by a screen from direct illumination from the lamp suspended in the sphere. The illumination received at this window is compared with a comparison lamp by means of a bench photometer external to the sphere.

The lamp under test is then replaced by a standard lamp. The illumination at the window due to this lamp is compared with the comparison lamp by means of the bench photometer. From the two measurements the luminous intensity of the test lamp may be calculated.

Mean hemispherical luminous intensity

This is the average luminous intensity measured in all directions below the horizontal.

Light meters

The lighting engineer's most useful instrument is the portable light meter, used to measure values of illumination on site. The latest type, and the most simple to use, consists of a photoelectric cell connected to a microammeter calibrated directly in lux (Fig. 9.6).

The photoelectric cell is responsive to light, and its electric response is proportional to the illumination received. The cell is provided with a shutter, so that the light may be cut off when the cell is not in use. The meter may read from 1 to 250 lux in two ranges controlled by switch or push-button.

A lighting survey of an area such as a street or room may be made quite simply by placing the light meter at predetermined points within the area and noting the readings.

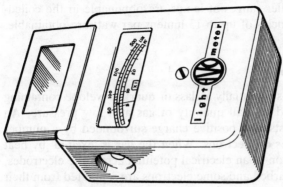

Fig. 9.6 Light meter

Incandescent filament lamps

When a body is heated it emits radiant energy in wave form.

As the temperature increases the wavelength decreases, until the waves are in the visible light range, from 4×10^{-5} to 7×10^{-5} cm. The light emitted is proportional to the 12th power of the absolute temperature, and this is the reason for running lamps at as high a temperature as possible.

Tungsten filament lamps

Drawn-wire tungsten filaments can be run in a vacuum at a temperature of 2150 °C. The efficiency of such a lamp is about 8 lumens per watt.

Gas-filled tungsten filament lamps.

If, after evacuation of air, the bulb is filled with an inert gas such as nitrogen or argon at a suitable pressure, the tungsten filament can safely be run at temperatures of 2400 to 2750 °C according to the size of the lamp. Owing to convection currents in the gas, loss of heat by convection tends to neutralize the increased luminous efficiency due to the increased temperature. By coiling the filament the loss by convection is reduced and an efficiency of 10 lumens per watt is obtained.

Coiled-coil filament lamps

The process of coiling the filament is carried a step further by coiling the coil. The smaller ranges of lamps are obtainable in the coiled-corn form. Efficiencies of up to 13 lumens per watt are obtainable.

Discharge lamps

A discharge tube is essentially a glass or quartz envelope containing two electrodes and a small quantity of gas at a low pressure. An atom of gas consists of a positive charge surrounded by a number of negative charges or electrons. When the gas is 'excited' by heat or by the application of an electrical potential across the electrodes, the atoms are disturbed and some electrons are separated from their atoms. An atom minus an electron has a net positive charge and is called a positive ion. With direct current the electrons move at high velocity towards the anode (or positive electrode), while the positive ions move more slowly towards the cathode (or negative electrode) and heat it by bombardment. The movement of the positive ions in one direction and of the electrons in the other constitutes the electric current (see Fig. 9.7). With alternating current the effect is the same, except that the flow of ions and electrons is reversed twice per cycle.

The velocity of the electrons depends upon the voltage applied, the higher the voltage the greater the velocity. The electrons collide with other gaseous atoms and remove further electrons. This is called 'ionization by collision'. Above a certain voltage, called the 'critical voltage', the action is cumulative and the tube will fail, unless the current is limited by the resistance or impedance of the external circuit.

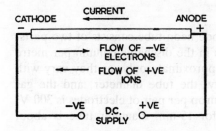

Fig. 9.7   Electron flow in discharge tube

As with the arc lamp, the resistance of the tube decreases as the current increases. A current-limiting device is therefore necessary, which is a series resistance with direct current or a current-limiting choke with alternating current. Discharge lamps are usually worked on alternating current.

The type of tube described above is called a 'cold-cathode' tube because the electrodes are not separately heated, and in general requires high voltages. The neon tube is an example of the cold-cathode tube.

Neon tubes

The popularity of high-voltage neon lighting arose almost entirely from its use in advertising, for signs, or in the decorative treatment of buildings, but later the lighting field became important. The neon tube, which is used in varying lengths up to about 9 m, may be bent into almost any desired shape during manufacture. It consists of a length of glass tubing containing two electrodes, normally cylindrical in shape, of iron, steel, or copper.

The true neon tube contains neon, but the term is now used also for tubes with fillings of other rare gases. The colourings available depend upon the particular gas or gas mixture used. Common colours are: Neon, orange-red; Helium, white; and Argon, blue. Further variations of colour are obtained by the use of coloured glass tube.

The diameters of the tube vary, and common sizes of 10, 15, 20, and 30 mm carry currents of 25, 35, 60, and 150 mA respectively. The neon tube requires series reactance for satisfactory operation, and this is generally effected by the use of a stray field transformer.

Voltage drop and power

The voltage drop in a cold-cathode neon tube consists of two parts, the electrode drop and the drop in the discharge column per metre of tube. The values given are approximate only, as they vary with the nature of the gas, its purity, the tube diameter, and the gas pressure. For neon, the voltage drop per pair of electrodes is 300 V, and the tube drop per metre for 15-mm diameter tube is about 400 V.

EXAMPLE

The neon outline of a sign contains 18 m of 15-mm tube in lengths of 3 m connected in series. Calculate the secondary voltage of the step-up transformer and its output in volt-amperes and watts. The power factor with corrective capacitor is 0.8. Find also the total lumens, assuming the lumens per watt to be 12.5.

Total voltage drop
= drop due to 18 m of tube + drop due to 6 pairs of electrodes
= $(18 \times 400) + (6 \times 300) = 7200 + 1800 = 9000$ V

Let current be 35 mA

Apparent power in volt-amperes $= 9000 \times 0.035 = 315$ VA
Power in watts $\qquad = 315 \times 0.8 = 252$ W
Given average lumens per watt $= 12.5$
Lumens emitted $= 252 \times 12.5 = 3150$ lumens

Hot-cathode discharge lamps

If one or more of the electrodes be heated, the electron emission is much increased and currents of 50 to 100 times greater can be carried. The electrode voltage drop is very much reduced, and the lamps will operate at ordinary voltages. Mercury vapour and sodium vapour are the most valuable fillings.

The vapour pressures of these gases at ordinary room temperatures are too low to permit the starting of a discharge, so these lamps contain, in addition, a few millimetres pressure of argon or neon which enables the discharge to start at a reasonable voltage. Current through the tube heats up the tube, increasing the vapour pressure, and the mercury or sodium discharge commences.

High-pressure mercury-vapour discharge lamps

These consist of a discharge envelope encased in an outer bulb of ordinary glass. The discharge envelope may be of hard glass or quartz. The space between the bulbs is partially or completely evacuated to prevent heat loss. The outer bulb absorbs harmful ultraviolet rays. The inner bulb contains argon and a certain quantity of mercury. When the discharge has attained a steady state, the mercury is completely vaporized. The electrodes consist of a filament of tungsten wire surrounding a stick of rare earths, but no independent filament heating is necessary.

An auxiliary electrode is used to start the discharge. It consists of fine wire and a high resistance R (see Figs. 9.8, 9.9). When the supply is switched on, full mains pressure is available across E_2 and E_3 and a glow discharge limited to a few milliamperes occurs. This enables the main discharge to commence.

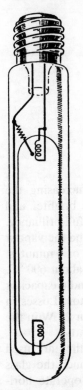

Fig. 9.8 High-pressure mercury-vapour discharge lamp

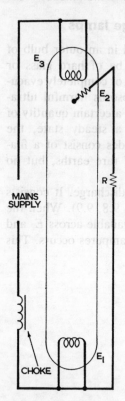

Fig. 9.9   High-pressure mercury-vapour lamp in diagram form

As the lamp warms up, mercury is vaporized, increasing the vapour pressure, and the luminous column becomes brighter and narrower. The lamp requires 4 or 5 minutes to attain full brilliancy. If the discharge is stopped, the lamp must cool down and the vapour pressure be reduced before it will restart. This takes 3 or 4 minutes. The temperature of operation inside the inner bulb is about 600 °C. The power factor is 0.6 with the special choke used, and a capacitor is connected across the mains to raise the power factor. Losses in the choke and capacitor amount to about 5 per cent, or 20 W in the 400-W lamp. The efficiency is about 40 lumens per watt.

These lamps must be operated vertically, since if they are used horizontally convection will cause the discharge to touch the glass bulb, which will fail. Lamps which are intended to operate horizontally are fitted with a magnetic device which will hold the luminous column central.

Sodium-vapour discharge lamps

These lamps have a characteristic yellow glow. They operate with low-pressure sodium vapour at a temperature of 300 °C. They have a neon filling with small globules of sodium. The neon initiates the discharge, which is taken up by the sodium vapour as the internal pressure increases. Because the efficiency falls off as the current density increases above a certain value, sodium lamps have low watt values compared with mercury-vapour lamps. Long discharge paths are necessary, and for this reason the discharge envelope is usually bent into U shape (Fig. 9.10). To insulate effectively the discharge envelope against loss of heat, the outer case is of the double-walled vacuum type. The lamp must be operated horizontally, or nearly so, to keep the sodium well spread out along the tube, although some small lamps may be operated vertically, lamp cap up. Several minutes are necessary after switching on before the discharge is at

Fig. 9.10   Sodium-vapour discharge lamp (GEC Ltd)

the full, but the lamps may be restarted immediately after a stoppage.

The sodium lamp is only suitable for alternating current, and therefore requires choke control. This requirement is met by operating the lamp from a stray field, step-up, tapped auto-transformer with an open-circuit secondary voltage of 470 to 480 V.

The uncorrected power factor is very low, about 0.3, and a capacitor must be used to raise the power factor to about 0.8. The efficiency is approximately 75 lumens per watt.

Fluorescent lamps

These are lamps coated with certain powders which glow when subjected to ultra-violet rays.

Low-pressure mercury-vapour lamps

In these the fluorescent powders are on the inside of the bulb itself. The electrodes are oxide-coated filaments which are heated independently during the starting period only and remain heated by virtue of the discharge. Although the voltage across the tube during operation is about 115 V only, a voltage much higher than the mains voltage is required to initiate the discharge. The circuit diagram for one tube is given in Fig. 9.11. When the mains switch S_2 is closed no discharge occurs, as the voltage is not high enough. If, however, the starting switch S_1 is also closed, the electrodes will heat up as they are then in series with the iron-cored choke across the mains. If S_1 is opened, the sudden reduction in current through the choke coil induces a momentary high voltage several times higher than the mains voltage, sufficient to start the discharge in the tube. The discharge itself then maintains the electrode temperature. The choke has a further function as a current-limiting or ballast impedance.

To obviate the need of manually operating two switches, the switch S_1 is made to work automatically. One type is the *glow type*, a small helium-filled discharge lamp with electrodes of bimetallic strip. When the main switch is closed, the potential across these electrodes causes a small glow discharge at a small current not sufficient to heat up the tube filaments. This discharge is enough, however, to heat the bimetallic strips of the switch itself, causing them to bend and make contact. A larger current then passes

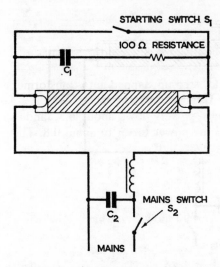

Fig. 9.11 Diagram of fluorescent lamp circuit

through the filaments, which heat up to red heat. Since the starting switch discharge is stopped when the contacts touch, the strips cool down and the contacts open. This opening of the contacts in series with the choke causes a momentary high voltage, which is sufficient to start the discharge in the main tube. The starting switch ceases to glow as the voltage is now too low.

Quick starting

A disadvantage of the fluorescent lamp as described above is the delay between switching on and steady illumination. Two main types of quick-start circuit have been devised, the pre-heat quick start and cold start.

Pre-heat quick starting

An earthed metal strip placed very near the tube will increase the voltage gradient at the cathode, and assist in starting ionization. Thus a tube provided with a very thin metal strip on the outside of the glass envelope will need a considerably reduced voltage. If in addition the electrode filaments are provided with sufficient current for very rapid heating, a 'quick start' is possible. Figure 9.12 shows

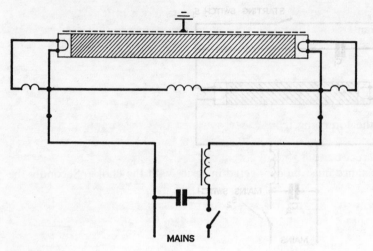

Fig. 9.12

this type of circuit. The primary winding of an auto-transformer is connected in parallel with the tube and receives practically full mains voltage. The filaments are provided with current from secondary tappings. When the lamp has started the transformer receives the normal tube voltage and the filament currents are correspondingly reduced.

Cold starting

This requires either the production of a circuit to give a much higher voltage kick when opened, or alternatively the direct application to the tube of a voltage approximately three times the normal tube voltage. The latter is obtained by the use of a step-up transformer. The tube electrodes are specially designed to withstand the higher voltages.

Twin-tube circuit

Other circuit arrangements of one or more fluorescent tubes are available, notably, twin-tube circuits. In one arrangement two single-tube circuits each without power-factor correction capacitors, are connected in parallel. One circuit remains at a low power factor of about 0.5 lagging, while the other circuit incorporates a series capacitor to give it a power factor of 0.5 leading. Thus the complete arrangement runs from the mains at unity power factor. Another advantage of this circuit arrangement is that owing to the phase difference between the currents in the two tubes, the stroboscopic effect on moving machinery is reduced.

Fluorescent lamps on direct current supply

In the foregoing it has been assumed that the supply to the fluorescent lamps is alternating current. If however, the available supply is direct current, modified circuits will be required. Firstly a ballast resistance must be connected in series with the choke. Secondly the starter switch must be specially designed for d.c. working, and thirdly a reversing switch must be connected into the circuit between the supply and the fitting. This is so that the current in the tube can be reversed at intervals to prevent migration of the mercury to one end of the tube.

Stroboscopic effect of discharge lamps

At the usual alternating current supply frequency of 50 Hz, a discharge lamp will be extinguished 100 times per second. Although this effect is seldom noticeable in normal conditions it can be a danger risk when running machinery attains certain critical speeds. For example, at 100 rev/s a revolving part could appear to be stationary. Some types of discharge lamp have an 'after-glow' which damps down stroboscopic effects. In other cases the use of lead-lag twin tubes is helpful, or in a large building adjacent rows of discharge lamps could be connected to different phases of the 3-phase supply.

A frequency flicker at half the above rate near the electrodes of fluorescent lamps is sometimes disturbing to the individual worker. In such a case this can be eliminated by the fitting of small plastic shields at each end of the tube, with a very small loss of light.

Emergency lighting

Premises licensed for public amusement are compelled by statute to have some form of secondary lighting to be used in case of emergency, i.e., the failure of the main lighting supply. Hospitals require emergency lighting in the operating theatre, and many large stores have a secondary lighting system installed.

Although two independent supplies from outside the building would in most cases be satisfactory, one supply for main lighting and the other for emergency lighting, the general method is to have a storage battery charged from the main supply, the battery to discharge into the secondary lighting system if the main supply and thus the main lighting fails. The change-over may be manual or automatic as the particular regulations and conditions demand. There are many systems in use, some giving normal voltage secondary lighting, and others low-voltage lighting. The battery may be charged either directly from direct current mains supply, or through a transformer and rectifier from alternating current mains. Other systems include a motor-generator set, the motor being alternating current or direct current as may be necessary.

If a battery is 'trickle-charged, that is, given a small constant charge to make up its open-circuit losses, it will keep in good condition, ready to deliver its full-rated output.

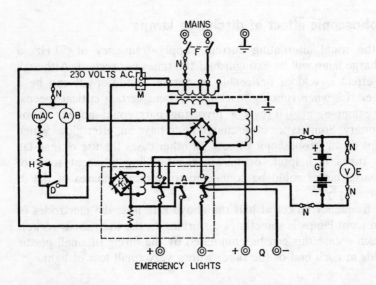

A	CONTACTOR	J	QUICK CHARGE CHOKE
B	AMMETER	K	COIL RECTIFIER
C	MILLIAMMETER	L	SELENIUM RECTIFIER FOR CHARGING
D	QUICK CHARGE SWITCH	M	MAINS SUPPLY TO CONTACTOR COIL
E	VOLTMETER		RECTIFIER
F	MAINS SWITCH	N	FUSES
G	BATTERY	P	RECTIFIER TRANSFORMER
H	TRICKLE CHARGE RESISTANCE	Q	'Q' SPECIAL CIRCUIT TERMINALS

Fig. 9.13 'Keepalite' emergency lighting (Chloride Batteries Ltd)

The 'Keepalite' emergency lighting system

Figure 9.13 gives the connections of one type of emergency lighting equipment. The 110-V battery is given a continuous trickle charge from the single-phase alternating current mains by means of a rectifier. In the event of failure of the mains supply, the contactor operates and connects the special emergency lights across the battery terminals. When the mains supply is restored the contactor changes over disconnecting the emergency lights and restoring the trickle charge to the battery. If required, a quick charge can be given to bring the battery up to its normal voltage.

Shades and reflectors

Specular reflection. A ray of light falling on a polished surface at an angle θ to the normal to the surface is reflected at the same angle (Fig. 9.14).

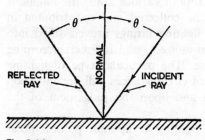

Fig. 9.14

Diffuse reflection

With matt surfaces the ray is broken up and reflected in all directions (Fig. 9.15).

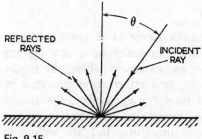

Fig. 9.15

Spread reflection

A polished surface, when roughened, will give a spread reflection in the general direction of specular reflection (Fig. 9.16).

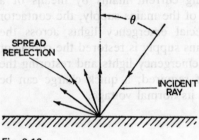

Fig. 9.16

Different materials and finishes give the above types of reflection in varying degrees.

By means of shades and reflectors of various kinds the luminous flux emitted from a lamp may be collected and redistributed in desired directions. Symmetrical lighting fittings are classified into direct, semi-direct, general, semi-indirect, and indirect, according to the light distribution they give. The particular type of lighting fitting to be used in a given set of circumstances will depend upon the lighting conditions required and upon the judgement of the lighting engineer.

Progressive manufacturers of lighting equipment publish polar curves in their literature to show the lighting effect of particular types of fittings. Figure 9.17 gives the characteristic curves of three types of recently developed 'Holophane' industrial units – extensive, medium, and focusing. The use of prismatic glass or acrylic ensures precise optical control and high efficiency.

Design of lighting schemes

It is not possible in a book of this character to do more than give the more simple principles employed in designing a new lighting scheme. For detailed information and extensive tables, the reader should consult specialist textbooks, and information provided by such bodies as the Illumination Engineering Society, the British Lighting Council, and the various manufacturers. These manufacturers not only supply descriptive literature, but are willing to give free advice on lighting problems.

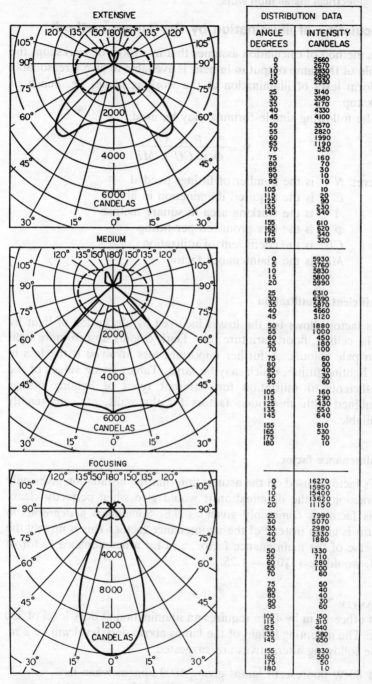

DISTRIBUTION DATA	
ANGLE DEGREES	INTENSITY CANDELAS

EXTENSIVE

ANGLE DEGREES	INTENSITY CANDELAS
0	2660
5	2670
10	2830
15	2880
20	2930
25	3140
30	3580
35	4170
40	4330
45	4100
50	3570
55	2820
60	1990
65	1190
70	520
75	160
80	70
85	30
90	10
95	10
105	10
115	20
125	90
135	230
145	340
155	610
165	620
175	340
185	320

MEDIUM

ANGLE DEGREES	INTENSITY CANDELAS
0	5930
5	5760
10	5830
15	5800
20	5990
25	6310
30	6390
35	5870
40	4660
45	3120
50	1880
55	1000
60	450
65	180
70	100
75	60
80	50
85	40
90	30
95	60
105	160
115	290
125	430
135	550
145	640
155	810
165	530
175	50
180	10

FOCUSING

ANGLE DEGREES	INTENSITY CANDELAS
0	16270
5	15950
10	15400
15	13620
20	11150
25	7990
30	5070
35	2980
40	2330
45	1880
50	1330
55	710
60	280
65	100
70	60
75	50
80	40
85	40
90	30
95	60
105	150
115	310
125	440
135	580
145	650
155	830
165	550
175	50
180	10

Fig. 9.17   Characteristic curves of 'Holophane' reflectors (Holophane Ltd)

Calculation of illumination by the 'lumen method'

This method of calculation assumes that in a room or workshop, the whole of the lamp output in lumens is available to give a reasonably uniform level of illumination at the working plane, i.e. bench or desk top.

The following simple formula may be used:

$$N = \frac{E \times A}{\Phi \times CU \times MF}$$

where: N is the number of fittings needed
       E is the required illumination in lux
       A is the working area in square metres
       Φ is the flux produced per fitting
      CU is the coefficient of utilization
      MF is the maintenance factor

Coefficient of utilization

This factor allows for the losses incurred by absorption of light by walls, ceiling, floor, furniture, etc. Dark colours absorb more light than pale colours. A further important loss involved is the loss in the lighting fitting, which may be large. Tables giving values of the coefficients of utilization for different types of lighting fittings combined with absorption factors for the walls, etc. are readily available.

Maintenance factor

This factor is used on the assumption that the installation gives only a fraction of the illumination it would give when perfectly clean. This factor is commonly given as 0.8. Sometimes a *depreciation factor* is given instead of the maintenance factor. This is merely the inverse of the maintenance factor, and for a maintenance factor of 0.8, would be $1.0/0.8 = 1.25$.

EXAMPLE 23
An office 18 m by 43 m requires an illumination at desk level of 330 lux. The mounting height of the lamps above desk level will be 2 m. The following alternatives are suggested:

(i) 80-W fluorescent lamps giving 4800 lumens when new,
(ii) 150-W tungsten-filament lamps giving 1950 lumens when new.

Calculate the number of lamps needed for each alternative assuming a coefficient of utilization of 0.6, and a maintenance factor of 0.85.

Use the formula $N = \dfrac{E \times A}{\Phi \times CU \times MF}$

then

(i) $N_F = \dfrac{330 \times (18 \times 43)}{4800 \times 0.6 \times 0.85} = 104$ lamps

(ii) $N_T = \dfrac{330 \times (18 \times 43)}{1950 \times 0.6 \times 0.85} = 257$ lamps

In spacing out the lamps, it is assumed that the distance between lamp centres in any row is approximately equal to the distance between adjacent rows. It is also assumed that the distances between the outside row and the wall, and between the end lamp and the end wall, is half the spacing distance.

If a scale plan of the office be drawn, the spacing of the lamps may be found either by calculation or by trial.

(i) There will be seven rows approximately 2.6 m apart, each of 15 fluorescent lamps, 105 lamps in all, the lamp centres in the rows being approximately 2.8 m apart.
(ii) There will be ten rows each of 26 tungsten lamps, a total of 260 lamps. The rows are approximately 1.8 m apart and the lamp centres are approximately 1.65 m apart.

The mounting height is the vertical distance between light fitting and the working level.

Spacing/mounting height ratio

In general, it is taken that this value should not be greater than 1.5. For the higher illumination values or for more uniform illumination the factor can be considerably less than 1.5.

Thus, in calculation (i) above:

$$\text{spacing/mounting height} = \frac{2.6}{2} \text{ and } \frac{2.8}{2}$$

$$= 1.3 \text{ and } 1.4$$

and in (ii):

$$\text{spacing/mounting height} = \frac{1.8}{2} \text{ and } \frac{1.65}{2}$$

$$= 0.9 \text{ and } 0.825$$

Point-to-point method

Different methods of calculation must be employed for open spaces, such as streets where the illumination is all received directly from the lamps, and for the interiors of buildings where the illumination at any given point is received partly (or in some cases wholly) by reflection from walls and ceilings.

In open spaces it is permissible to calculate the illumination on the basis of the polar distribution of the light from the lamps and their reflectors, using a point-to-point method and employing the cosine law.

EXAMPLE

Two lamps which have equal luminous intensities of 1000 candelas in all directions below the horizontal are mounted 6 m above ground level and 20 m apart.

Lamps No. 1 and No. 2 will give a similar distribution of illumination (Figs. 9.18 and 9.19). The method is very tedious when a number of lamps are spaced over an area. Figure 9.20 is a graph of the total illumination along the line joining the two lamps. Although the illumination distribution may be satisfactory for area

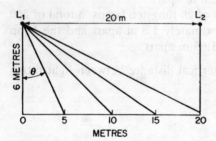

Fig. 9.18

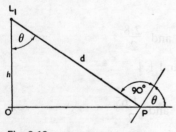

Fig. 9.19

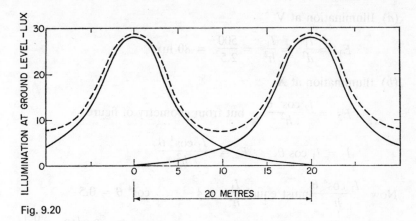

Fig. 9.20

lighting in some conditions, it would not be suitable for internal lighting. Either the spacing must be reduced, or some form of directional reflector must be installed.

EXAMPLE

An incandescent-filament lamp suspended 2.5 m above a work bench is fitted with a reflector to give a polar curve which has the shape of a circle with the circumference passing through the centre of the light source and giving an intensity of 500 cd vertically below the lamp.

(a) Calculate the illumination on the bench vertically below the lamp.
(b) Find the position along the bench where the illumination will be half the value found in (a).

Refer to Fig. 9.21.

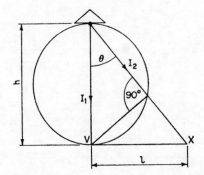

Fig. 9.21

(a) Illumination at V

$$E_1 = \frac{I_1}{d^2} = \frac{I_1}{h^2} = \frac{500}{2.5^2} = 80 \text{ lux}$$

(b) Illumination at X

$$E_2 = \frac{I_2 \cos^3 \theta}{h^2}, \quad \text{but from geometry of figure}$$

$$I_2 = I_1 \cos \theta, \quad \therefore E_2 = \frac{I_1 \cos^4 \theta}{h^2}$$

Now $\dfrac{I_1 \cos^4 \theta}{h^2}$ must equal $\dfrac{I_1}{h^2} \times \frac{1}{2}$ $\quad \therefore \cos^4 \theta = 0.5$

$$\therefore \cos \theta = 0.8410$$

$$\text{and } \tan \theta = 0.6445$$

$\tan \theta = \dfrac{l}{h}$ $\quad \therefore l = h \tan \theta = 2.5 \times 0.6645 = 1.61$ metres

10 Electric heating

Heat is a form of energy, and one of the valuable features of electricity is the ease with which electrical energy can be converted into heat energy.

There are two qualities of heat which can be measured: *intensity* and *quantity*.

Temperature is a measure of the *intensity* of heat, and is recorded in the lower ranges by thermometer. Pyrometers are used for the higher temperatures. Until recently the common scales by which temperature has been measured in Britain and in some other countries have been Fahrenheit (°F) for everyday use and centigrade (°C) for scientific purposes. The freezing and boiling points of water are respectively 32 °F and 212 °F, or 0 °C and 100 °C. Thus a range of 180 °F covers a range of 100 °C. The conversion of a temperature reading from one system to the other can be effected by using the equations below:

Temperature in Fahrenheit degrees
$$= \tfrac{9}{5} \text{ (temperature in centigrade degrees)} + 32$$

or, Temperature in centigrade degrees
$$= \tfrac{5}{9} \text{ (temperature in Fahrenheit degrees} - 32)$$

Since SI units are now to be used for all purposes in Britain, the temperature scale to be used is the Celsius scale. The name centigrade has been replaced by the name Celsius, and 1 degree centigrade becomes 1 degree Celsius. The *absolute* scale of temperature is in kelvin (symbol K), where,

temperature in kelvin = temperature in °C + 273.15

The symbols K or °C are to be used for points on the temperature scale and also for temperature intervals. In this book all references and calculations concerning temperature deal with the Celsius scale.

Quantity of heat

Heat is a form of energy and therefore the same unit is used for quantity of heat as for other forms of energy, namely the joule.

It has been found experimentally that the quantity of heat required to raise the temperature of 1 kg of water 1 °C is 4187 J. Therefore the quantity of heat required to raise the temperature of a mass of water is given by

$$Q = 4187 \, mt$$

where Q is quantity of heat in joules
        m is mass in kilogrammes
        t is temperature rise in degrees Celsius.

EXAMPLE

Find the quantity of heat needed to raise the temperature of 20 kg of water from 30 °C to 330 °C.

$$
\begin{aligned}
Q &= 4187 \, mt \\
  &= 4187 \times 20 \times (330 - 30) \\
  &= 4187 \times 20 \times 300 \\
  &= 25\,100\,000 \text{ J}
\end{aligned}
$$

Specific heat capacity

The heat required to raise the temperature of 1 kilogramme of a material through 1 Kelvin (or 1 degree Celsius) is called the Specific Heat Capacity (SHC). For water the specific heat capacity is 4187 J/kg K. Thus if c is the specific heat capacity, then

$$Q = c \, m \, t$$

Table 10.1

Substance	Specific heat capacity (J/kg K)	Density (kg/m³)
Water	4187	1000
Air	1010	1.292
Aluminium	946	2720
Copper	390	8930
Iron	452	7870
Lead	126	11 340
Mercury	1394	13 570
Nickel	444	8860
Silver	234	10 490
Tin	225	7300
Oil (transformer)	2140	898

A number of values of specific heat capacity for various substances are given in Table 10.1; e.g. SHC for copper = 390 J/kg K. Thus in the example above, if the material were copper instead of water,

$$Q = 390 \times 20 \times 300 = 2\ 340\ 000 \text{ J}$$

EXAMPLE

Calculate the heat energy in joules required to raise the temperature of:

(a) 4.5 litres of water from 15 °C to 100 °C. (The mass of 1 litre of water is 1 kg)

(b) 0.028 m³ of copper from 0 °C to 60 °C

(c) 23 kg of iron from 10 °C to 80 °C

(d) 34 m³ of air from 10 °C to 18 °C

(e) 7 kg of lead from 0 °C to 90 °C

(a) Joules (J) required
$$= 4187 \times 4.5 \times (100 - 15)$$
$$= 4187 \times 4.5 \times 85 \qquad = 1\ 602\ 000 \text{ J}$$

(b) J $= 390 \times (0.028 \times 8930) \times 60$
$$= 390 \times 250 \times 60 \qquad = 5\ 850\ 000 \text{ J}$$

(c) J $= 452 \times 23 \times (80 - 10)$
$$= 452 \times 23 \times 70 \qquad = \quad 728\ 000 \text{ J}$$

(d) J $= 1010 \times (34 \times 1.292) \times (18 - 10)$
$$= 1010 \times 43.9 \times 8 \qquad = \quad 355\ 000 \text{ J}$$

(e) J $= 126 \times 7 \times 90 \qquad = \quad 79\ 380 \text{ J}$

Transfer of heat

The different ways by which heat is transferred are conduction, convection, and radiation.

Conduction

This is the transfer of heat through a substance, from one part to another, or between two substances in contact.

Convection

The air in contact with a heated radiator element in a room receives heat from contact with the element. The heated air expands and rises, cold air flowing in to take its place. Thus there is a constant flow of air upwards across the heated element. This process is called convection. These convection currents give up some of their heat to the colder parts of the room. The room and its contents are gradually heated by this means. A similar action takes place in an electric water-heater, convection currents causing a continuous flow of water to pass upwards across the immersed heater element, with the result that the whole of the water in the tank becomes heated.

The quantity of heat absorbed from the heater by convection depends chiefly upon the temperature of the heater above the surrounding air or water and upon the size of the surface area of the heater. It also depends partly on the position of the heater. The quality of the heater surface, polished or matt, has no effect.

Radiation

Earth is warmed by energy received from the sun; this energy travels through the vacuum between the two by what is called radiation. All hot objects emit heat by radiation. Radiation, like light, is mainly reflected from a bright polished surface, but almost wholly absorbed by a dark matt surface.

Conversion of electricity to heat

When an electric current passes through a conductor the power spent in the conductor in watts is given by the formula, P watts $= VI$, where V is the voltage drop in the conductor and I is the current in amperes.

Since, by Ohm's law, $V = IR$, then $P = I^2R$, where R is the resistance of the conductor in ohms.

This energy is converted into heat which raises the temperature of the conductor. As the conductor gets hot, it gives off some of its heat to the surrounding atmosphere or substance. The higher the temperature of the conductor the faster it will loose heat. Thus the temperature of the conductor will rise until heat is lost at the same rate as it is generated, when the temperature will remain steady.

Methods of heating rooms

The amount of heat required for personal comfort in an enclosed space such as a room or office varies very considerably with the following:

- Number of changes of air per hour
- Area of windows
- The situation of the walls, external or internal (most rooms have at least one external wall)
- The exposure of the ceiling (open to the roof or with heated bedroom over)
- Material of which walls, floors, and ceilings are composed
- The outside air temperature.

A rough approximation which may be used in calculating the heat required, under normal English conditions, is 53 W per m^3 of air space.

The quantity of heat developed by the heater may be reduced when the desired temperature is reached.

Types of electric heater

These include:

1. Open-type radiators or electric fires
2. Tubular heaters
3. Convector heaters
4. Panel heaters
5. Thermal storage systems
6. Floor-warming arrangements
7. Night-storage heaters

Electric fires or radiators

Such heaters consist of spiral resistances wound on fireclay formers and working at a luminous temperature of from 1400° to 1600 °C. About 50 to 60 per cent of the heat emitted is in the form of radiant heat, while the remainder is spent in warming the air convection currents. The usual sizes are from 600 W to 3 kW. They may be either portable or fixed.

Radiant fires are controlled by ordinary switches and are not suitable for thermostatic control.

Tubular heaters

These are steel tubes about 50 mm diameter containing resistance elements arranged on mica, fireclay, or porcelain formers. The normal loading is 200 W per metre length. The tube temperature

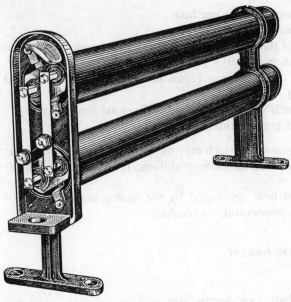

Fig. 10.1   Tubular heating (GEC Ltd)

is about 200 °C, which is attained in about 20 minutes. Tubular heaters may be fixed as single tubes up to about 6 m in length or in vertical banks. The usual position for fixing the single tube is in the angle of the wall and floor. These tubes may be considered as convector heaters, as very little of the heat is given up to the room as radiation. Air flows across them at low velocities, rises, and circulates through the room. The tubes are sufficiently hot to burn fabrics, and should be protected where necessary, especially in nurseries. Figure 10.1 shows a 2-tube bank with connection block. Tubular heaters are specially suitable for thermostatic control, by which means they are switched on and off automatically as the room temperature falls below or rises above the desired value. When fixed at the base of a cold wall or underneath a window they prevent cold down-draughts, which cause cold feet to the occupants of the room.

Convector heaters

These consist of wound resistance elements contained within a sheet-metal case with inlet and outlet openings or louvres at the bottom and top respectively. The front and top may be of moulded

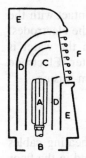

Fig. 10.2   Cross-section of convector heater

plastic material for the sake of pleasing appearance. These heaters may be either portable or fixed, and may be handled without fear of burns. The action may easily be understood by reference to Fig. 10.2 which shows cold air entering the bottom of the heater, passing over the heater elements, and leaving the heater by the top opening as warm air.

A very satisfactory method of heating a room is to use convector heaters to warm the room to about 14 °C, together with a small radiant fire for local heating.

Panel heaters

These heaters are made in the form of flat panels in which resistance wire is embedded. Low-temperature panels operating at temperatures ranging from 30 °C to 65 °C may be fixed in the walls or ceiling of a room, and plastered over. High-temperature panels up to 340 °C are used on walls and ceilings but are not fixed in the wall surface. They are sometimes used suspended from the ceiling. The greater part of the heat entering the room from these panels is in radiant form. Loadings are of the order of 430 W per m^2 for the low-temperature panels and 6400 W per m^2 for the high-temperature panels.

Thermal storage system

Roughly this system is similar to the coal, coke, or oil fired central-heating system, except that the heat is supplied to the storage cylinder electrically. There are two types of heater: the immersion heater used on voltages up to about 650, and the electrode type used at higher voltages. The electrode type is only suitable for use on alternating current. The electrodes are in direct contact with the water, and current passes through the water between the electrodes. The energy expended in passing the current through the water is turned into heat and raises the temperature of the water. Electrode heaters may be used either for water-heating or steam-raising.

Floor warming

A floor-warming system is in essence a number of heating cables embedded in the floor of a room or otherwise contained in the floor space. Their function is to provide slow heat to the concrete or other floor material. The heat from the floor is employed in warming the air in the room, and in warming the walls and the furniture. The electrical power intake is controlled by a thermostat fixed on the room wall in a suitable position.

The heater cables are constructed of alloys of nickel, chromium, iron or other resistance materials, and are insulated with asbestos, mineral insulation, butyl rubber, or other heat-resisting insulating materials. Mechanical protection can be by lead, copper, aluminium, or p.v.c. sheathing.

To obtain an even diffusion of heat, the heating wires should be arranged to reach almost from wall to wall at regular intervals of say 100 mm. Loadings of from 110 to 160 W per m^2 of floor are generally satisfactory.

The heating cables may be connected to the normal mains supply whereby the power input can be used at any time, but will be charged at normal heating rates. A common alternative is to arrange for an 'off-peak' supply, in which case the hours of supply are limited mainly to night-time hours; the tariff is much cheaper than the normal tariff. With this type of space-heating, it may be desirable to install one or more convectors or radiant heaters for use towards the end of the floor-warming output cycle.

There are a number of different systems available of which details may be obtained from the makers. Many consumers prefer the draw-in systems, in which cables may be taken out for repair or replacement.

Night storage heaters

The use of this type of heater for domestic purposes is increasing rapidly, particularly since it is linked with 'off-peak' supplies and consequently cheaper running costs.

The heater consists of resistance elements encased in ceramic insulating tubes, situated between blocks of heat storage material. The whole is enclosed in a metal casing. The charging arrangements are similar to those of the floor-warming systems already described. The special 'off-peak' circuit includes a time switch which controls the automatic charging of the heater, giving a long night charge and a short midday boosting charge.

A wide variety of heaters are available, some of which have a simple uncontrolled heat output, and others which have the heat output controlled by flaps or dampers and/or fans which allow more or less air to flow through the heater case. The dampers and fans may be manually controlled, or may be automatically controlled by thermostatic arrangement.

Calculation of loading required for an electric heating installation

Table 10.2 Table for calculation of loading required for rooms of up to approximately 280 m³ capacity

Number of exposed walls	Ceiling	Floor	Watts per m³ (assuming window space to be 25% of floor area and 16.5 °C maximum temperature rise over outside temperature)			
			215-mm walls		300-mm walls	
			Intermittent heating	Continuous heating	Intermittent heating	Continuous heating
4	Warm	Warm	57.6	43.1	51.3	38.5
	Cold	Warm	67.8	51.0	61.5	46.4
	Warm	Cold	61.8	46.7	55.8	42.1
	Cold	Cold	72.4	54.4	65.4	49.8
3	Warm	Warm	49.1	36.8	44.2	33.5
	Cold	Warm	59.3	44.5	54.8	41.0
	Warm	Cold	53.7	40.3	49.5	37.1
	Cold	Cold	63.9	48.1	56.9	42.8
2	Warm	Warm	40.3	30.6	37.5	28.5
	Cold	Warm	51.0	38.1	48.0	36.1
	Warm	Cold	45.2	34.2	42.9	31.8
	Cold	Cold	55.8	41.7	53.0	39.6
1	Warm	Warm	31.8	24.2	31.0	23.5
	Cold	Warm	42.1	31.7	41.4	31.4
	Warm	Cold	36.8	27.8	35.7	27.0
	Cold	Cold	47.0	35.3	46.0	34.9

The table and calculation which follow are printed by permission of the General Electric Co Ltd.

Table 10.2, with example below, is given to enable a quick and approximate estimate to be made. Results will probably be accurate to within 10 per cent either way.

EXAMPLE

Suppose it is desired to heat a room 6 m by 4.5 m, by 3 m high. The room has three exposed 300-mm walls, a warm floor (heated from the room below, which is itself already heated), and a cold ceiling (there being an unheated attic or room above). The heating is required to be intermittent, that is, only at certain periods of the day or night.

The cubic capacity of the room is $6 \times 4.5 \times 3 = 81$ m^3. Taking the various factors into account, we find that, reading from the table, 54.8 W per m^3 would be required. Therefore this loading is $54.8 \times 81 = 4440$ W.

Room heating calculations

More detailed calculations of the heating of buildings can be made by making use of standard tables of heat loss. The heat transfer through a wall, for instance, depends upon the materials, make-up, and thickness of the wall, and upon the temperature difference between the two sides of the wall. Heat will always flow from a point of higher temperature to one of lower temperature.

Heat transfer coefficients

The heat transfer coefficient or U value of a material is given as the amount of heat energy conducted through unit area of the material in unit time with unit air-temperature difference between the two sides. The SI unit is W/m^2 °C. Tables giving U values for building materials singly or in combination are available in specialist literature and textbooks.

If values are extracted from the tables for walls, ceiling, floors, doors, and windows, the heat losses may be calculated. In many instances in offices or houses, the heat losses through a particular partition may be very small, since there may be very little temperature difference between the two sides.

The other heat used in a room is that needed to heat the air in the room and thus to maintain the comfort of the occupants. The number of complete changes of air per hour must be estimated. The usual value taken for calculation purposes is 2.

A drawing office, 15 m by 10 m with ceiling height 4 m, has a door area of 5 m^2 and a window area of 45 m^2. The office is to have electric heating sufficient to maintain an average inside temperature of 19 °C when the temperature outside walls, ceiling, and floor is 0 °C.

Calculate the kW rating of the heaters assuming *two* complete changes of air per hour.

Specific heat capacity of air = 1010 J/kg °C
Density of air = 1.292 kg/m^3

Heat transfer coefficients (U values)

Floor – wood blocks on concrete   0.85 W/m^2 °C
Ceiling – asphalt, concrete, polystyrene 0.99   "
Walls – double brick and plaster   1.76   "
Doors – wood   2.84   "
Windows – double-glazed   2.90   "

Volume of room   = 15 × 10 × 4   = 600 m^3
Temperature rise   = 19 °C
Area of floor   = 15 × 10   = 150 m^2
Area of ceiling = 15 × 10   = 150 m^3
Area of walls   = (50 × 4) − (45 + 5)
                 = 200 − 50   = 150 m^2

Heat loss through structure

P = area(m^2) × U value(W/m^2 °C) × temp. rise (°C)

Part	Area	U Value	Temp. rise	Heat loss
Floor	150	0.85	19	2420
Ceiling	150	0.99	19	2820
Walls	150	1.76	19	5020
Doors	5	2.84	19	270
Windows	45	2.90	19	2480
				13 010 W

Heat loss to air

J = Spec. heat capacity × mass × °C × 2 (changes)
= 1010 × (600 × 1.292) × 19 × 2 = 29 760 000

$$P = \frac{29\,760\,000}{3600} = 8720\text{ W}$$

Total heat loss = 13 010 + 8270 = 21 280 W
  Thus, kW rating of heaters = 21.3 kW

Thermal insulation

To reduce heat loss through the fabric of a building, a variety of methods and heat (thermal) insulating materials may be used. Within the existing Building Act, any new building having a cavity wall structure must have this cavity filled, as building progresses, with thick panels of fibre glass.

It is commonplace to insulate any roof void or loft, by either filling the space between the ceiling/roof joist with granulated vermiculite or expanded polystrene chippings, or by laying rolls of fibre glass which are manufactured to fit between the joists.

Fibre glass in a quilt or blanket like form of up to 2 metres in width may be laid across the tops of the ceiling joists. This method has the benefit of not only ease of installation, but also traps the air between the ceiling and the fibre glass quilt. Air trapped in this way also acts as an insulation, thus reducing heat loss even more.

Windows and glazed doors which are 'double' glazed also use this principle of trapping air between the two sheets of glass held within the window or door frame. Colder climates like Sweden or Canada are now using triple glazing, whilst the thickness of fibre-glass quilting has increased from 25 mm to 150–200 mm.

Heating water

Immersion heaters

The complete unit consists of a copper container insulated on the outside with a thick layer of heat-insulating material, e.g. granulated cork. The whole is then enclosed in a thin steel casing. The cold water inlet feeds the bottom of the tank, and the hot water is drawn

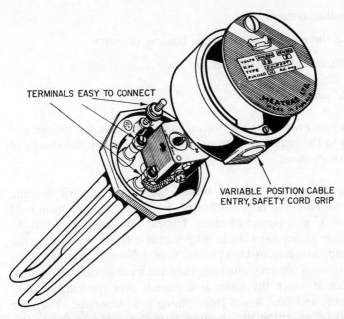

TERMINALS EASY TO CONNECT

VARIABLE POSITION CABLE
ENTRY, SAFETY CORD GRIP

Fig. 10.3   Modern form of immersion heater with rod-type thermostat
(Heatrae Ltd)

from the top. The water is heated by means of an immersed heating
element. This is a tube or tubes in U-shape which encloses the
resistance element. The one illustrated in Fig. 10.3 is of 80/20 nickel
chrome wire embedded in magnesium oxide in a seamless copper
tube. The tube is welded to the brass connection plate. A thermostat is connected in series with the element, to break the circuit
when the water reaches the required temperature. The rod type of
thermostat shown consists of an outer copper tube closed at one
end, in which is fixed an alloy rod of negligible coefficient of expansion. The rod and tube are fixed solidly together at the end remote
from the connection plate. Thus, as the water heats up the tube
expands drawing the rod away from the micro-switch mechanism at
the connection plate and opens the switch contacts. The micro-
switch contacts are sometimes provided with a mechanical or a
magnetic snap action to prevent flutter at the critical temperature.
For direct current work a small capacitor may be connected across
the switch contacts to prevent arcing.

Water heating systems

There are three systems of electric heating of water:

- Instantaneous heaters or electric geysers
- Local storage
- Central storage

Instantaneous heaters.   Cold water passes through the heater and is heated to the required temperature by contact with the sheath of the electric element.

Local storage.   This type of heater has a low electric loading sufficient to raise the temperature of the water contents from cold to, say, 80 °C in a period of about 1 hour. For example, a 500-W, 7-litre water heater working at 90 per cent efficiency will raise the water temperature from 16 °C to 80 °C in 1 hour 10 minutes.

These heaters are very efficient; they are fixed as near as possible to the spot at which the water is required, over the kitchen sink, for instance, and heat losses from piping are eliminated. They are controlled by an automatic thermostat incorporated in the apparatus, and by a local single-pole switch. They are usually of the 'free-outlet' type.

Central storage.   This is the electrical counterpart of the normal domestic coal or coke-fired hot-water installation. The water is heated by means of an immersion heater fixed in the hot-water cylinder. The heater may be controlled by a thermostat which will switch the heater into circuit when water is drawn off and the water temperature falls, and switch off again when the maximum water temperature is again reached.

11   Direct current machines

The direct current generator

The direct current generator comprises an electromagnet with one or more pairs of poles, an armature winding consisting of a number of conductors in series or in parallel, an armature core upon which the winding is mounted, and a commutator which has the effect of changing the alternating e.m.f. generated in the winding into a direct or continuous e.m.f.

If a conductor is caused to move in a magnetic field so as to cut the lines of force, an e.m.f. is generated in the conductor, the value of which is proportional to the strength of the field, to the length of the conductor in the field, and to the speed of the conductor through the field. Expressed in another way, the e.m.f. is proportional to the rate of cutting lines of force. Figure 11.1 shows the direction of the e.m.f. generated in a conductor moving across a magnetic field, in accordance with Fleming's right-hand rule. If the thumb, forefinger and second finger of the right hand are held at right angles to each other, so that the thumb indicates the direction of motion of the conductor, and the forefinger indicates the

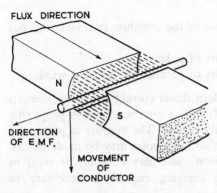

FLUX DIRECTION

DIRECTION
OF E.M.F.

MOVEMENT
OF
CONDUCTOR

Fig. 11.1   Illustrating generation of e.m.f.

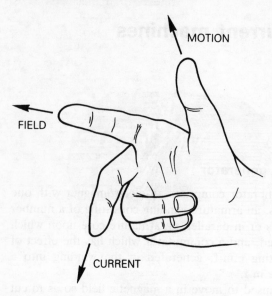

Fig. 11.2 Fleming's right-hand rule

direction of the magnetic field flux, then the second finger will indicate the direction of the induced e.m.f. as shown in Fig. 11.2.

The general formula for the e.m.f. of a direct current generator is,

$$E = \frac{\phi Z n p}{a} \text{ volts}$$

where ϕ is the flux per pole, in webers

Z is the total number of conductors on the armature winding

n is the speed of rotation of the armature in revolutions per second

p is the number of pairs of poles

a is the number of pairs of parallel paths in the winding

The electromagnet of a modern direct current generator consists of a circular yoke of cast steel, with a number of poles projecting inwardly, of polarity N and S alternately. The number of poles may be two, four, or any even number. The poles may be made of cast iron, cast steel, or wrought iron, and may be solid or built of laminations. A field winding carrying current is necessary to

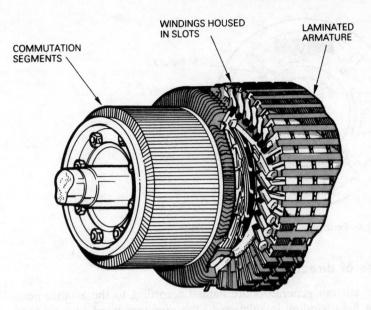

COMMUTATION
SEGMENTS

WINDINGS HOUSED
IN SLOTS

LAMINATED
ARMATURE

Fig. 11.3   Typical of commutation and armature

magnetize the poles, and this winding is wound on formers and
mounted on the pole pieces. The armature core consists of iron
stampings bolted up together on the shaft of the machine and so
arranged as to allow for efficient ventilation. Stampings are used
instead of a solid core to cut down eddy current losses. The armature
conductors, in which are generated the e.m.f., are wedged into
insulated slots in the surface of the armature core. The whole
armature is caused to rotate by mechanical power. The commutator,
which rotates with the armature, is built up of a number of copper
commutator bars insulated from each other and from the shaft. The
ends of the armature conductors are soldered or welded to the risers
of the commutator bars. Carbon brushes are arranged to bear upon
the commutator. To assist in sparkless commutation, small auxiliary
poles called interpoles or commutating poles are often built into the
machine. The open type of enclosure is usually satisfactory for
direct current generators as there is normally little need for special
mechanical protection. Figure 11.4 shows a 4-pole machine with
interpoles, with field winding on one pole. Other poles and interpoles are shown without windings.

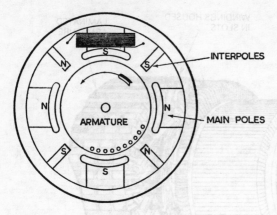

Fig. 11.4   Four-pole direct current generator

Types of direct current generator

Direct current generators are named according to the arrangement of the field winding in relation to the armature winding. The four chief types are, respectively: separately excited, series-wound, shunt-wound, and compound-wound. The compound winding may be 'short shunt' or 'long shunt' (see Fig. 11.5). A fifth type has a permanent magnet and no field winding.

Separately excited generator

This machine is seldom used except for special purposes. The curve of terminal voltage against current output is given in Fig. 11.5 (*a*). The slight fall in terminal voltage with increasing load is mainly due to the armature resistance voltage drop.

Series-wound generator

This is the type of generator used as a 'booster' on long traction cables. Since the field winding is in series with the armature, the machine will only generate its rated voltage if the external circuit is closed. Figure 11.5 (*b*) shows the curve of voltage against current output.

Shunt-wound generator

This is the most common type of generator. The curve of terminal

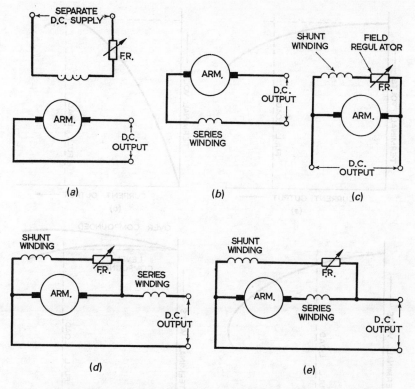

Fig. 11.5 Direct current generators (a) Separately excited (b) Series wound (c) Shunt-wound (d) Short shunt compound-wound (e) Long shunt compound wound

voltage against current output shows a greater fall in voltage than that of the separately excited generator. As the terminal voltage falls off due to armature resistance, the voltage across the field winding is reduced, thus decreasing the field current. This results in less field flux in the magnetic circuit, further reducing the e.m.f. The double effect gives a curve, as shown in Fig. 11.6 (c).

Compound-wound generator

This machine has a series field winding in addition to the shunt field winding. This series winding is usually arranged to assist the shunt winding. Thus the series winding tends to raise the e.m.f. generated as the load current increases. If the winding is strong enough this increase may compensate or more than compensate for the drop of

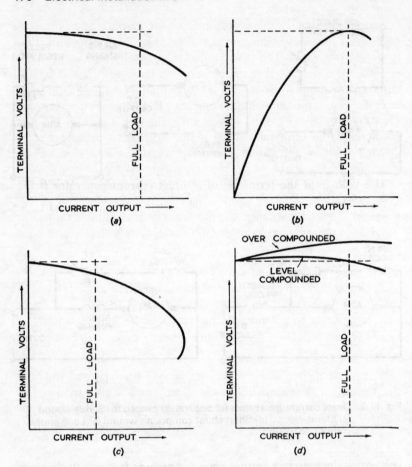

Fig. 11.6 (a) to (d) Curves showing variation of terminal voltage with current output

voltage in the armature, brushes, and series winding. If the drop is exactly compensated; the terminal voltage remains constant, and the machine is called 'level' compounded. If it is over-compensated the voltage rises, and the machine is said to be 'over-compounded' (see Fig. 11.6 (d)).

Field-regulating resistance

If an adjustable resistance is connected in the shunt circuit of a generator in series with the shunt winding, the field current may be increased or decreased by cutting out or putting in resistance. Increasing the field current increases the field flux, and thus increases the e.m.f. The field regulator, normally hand-controlled, is used, therefore, to increase the e.m.f. as the load becomes larger, sufficient to compensate for the greater armature voltage drop, thus keeping the terminal voltage constant. If the generator supplies current through fairly long cables, the field regulator may be used to increase the terminal voltage above normal to compensate for voltage drop in the cables. The field regulator is shown in the diagrams of Fig. 11.5.

The voltage at the terminals of a direct current generator is the generated e.m.f. less the drop in volts in the machine due to the armature internal resistance and the brush contact voltage drop. For any direct current generator the terminal voltage $V = E - (I_a r_a + I_a r_a + \text{brush contact drop})$,

where E is the generated e.m.f.,

$\quad I_a$ is the current in the armature

$\quad r_a$ is the resistance of the armature

$\quad r_s$ is the resistance of the series field winding, including commutating poles, if any.

The brush contact voltage drop is about 2 V irrespective of the current passing. In a shunt generator there is no series winding, and the quantity $I_a r_s$ is zero unless there are commutating poles. The formula shows that the terminal voltage decreases as the generator loading increases, owing to the increase in the armature voltage drop.

EXAMPLE

A shunt generator supplies current at a terminal pressure of 240 V to a lighting load of 25 kW. The resistance of the field winding is 80 Ω and the armature resistance is 0.16 Ω.

Assuming a total voltage drop of 2 V due to brush contact, calculate:

(a) the armature current
(b) the generated e.m.f.

(a) Load current $= \dfrac{W}{V} = \dfrac{25\ 000}{240} = 104$ A

$\quad$ Field current $= \dfrac{V}{R} = \dfrac{240}{80} = 3$ A

Therefore, armature current $= 104 + 3 = 107$ A

(b) Generated e.m.f. $= V + I_a r_a + 2$
$= 240 + (0.16 \times 107) + 2$
$= 240 + 17 + 2 = 259$ V

Armature reaction, which cannot be further discussed here, has the effect of reducing the terminal voltage in all the direct current generators here described.

Direct current motors

If a conductor carrying a direct current lies in a magnetic field at right angles to the lines of force, the reaction between the field and the field due to the conductor sets up a mechanical force which tends to cause movement of the conductor. The direction of movement depends upon the direction of the field, and the direction of the current in the conductor. In Fig. 11.7 the movement of the conductor would be downwards, in accordance with Fleming's left-hand rule. If the thumb, first and second finger of the left hand are extended and at right angles to one another, as shown in Fig. 11.8. The first finger points to the direction of magnetic field, with the second finger representing the direction of the current flow, the thumb position will give the direction of motion for the conductor due to the force acting on it. If *either* the current or the field were reversed, the direction of movement would be upwards. If *both* were reversed the direction of movement would still be downwards. The mechanical force on the conductor is proportional to the strength of the magnetic field, to the current in the conductor, and to the length of the conductor under the influence of the field.

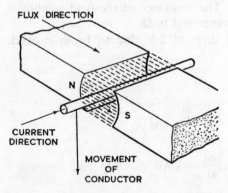

Fig. 11.7   Illustrating mechanical force on a conductor

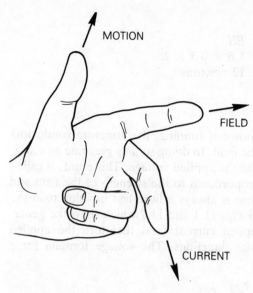

Fig. 11.8   Fleming's left-hand rule

By formula, force on conductor = BlI newtons,

where: B is the flux density in webers per square metre (Wb/m^2)
        l is the length of the conductor in metres
        I is the current in the conductor in amperes

(*Note*: the SI unit of flux density is the tesla, symbol T where
I T = 1 Wb/m^2)

If the field magnets of a direct current generator is excited with
direct current and direct current is supplied to the armature, the
machine will be a direct motor. The armature will rotate, converting
electrical power into mechanical power. A direct current motor has
essentially the same construction as a direct current generator,
although in general the external appearances differ owing to the
different conditions under which direct current motors are expected
to work.

EXAMPLE

A conductor 300 mm long lies at right-angles to a magnetic field of
density 1.6 T, and carries a current of 24 A.
    Calculate the force on the conductor.

By formula:

$$\text{Force in newtons} = BlI$$
$$= 1.6 \times 0.3 \times 25$$
$$= 12 \text{ newtons}$$

Back e.m.f. of a motor

When a direct current motor is running, the armature conductors cut the lines of force of the field. In doing so they generate an e.m.f. which acts in opposition to the applied voltage. This e.m.f. is called the 'back e.m.f.'. It is proportional to the strength of the field and to the armature speed, but is always a little less in value than the applied voltage. Refer to Figs 11.1 and 11.7 which show the generated e.m.f. and the applied current (and therefore the applied voltage) to be in opposite directions. The voltage formula for a direct current motor is

$$V = E_b + I_a r_a$$

where: V is the applied voltage
           F_t is the back e.m.f.
           $I_a r_a$ is the voltage drop in the armature

For any particular motor with a fixed applied e.m.f. the back e.m.f. will have a fixed value for a particular value of armature current irrespective of the motor speed. From the e.m.f. formula already given on p. 172, the back e.m.f. E is proportional to the flux and to the speed:

$$E \propto \phi n$$

where ϕ is the flux per pole and n is the speed in rev/s.

Hence, $\phi \propto \dfrac{E}{n}$, or $n \propto \dfrac{E}{\phi}$

Thus, for any particular loading, flux and speed are inversely proportional to each other. Thus, if the field flux is decreased by reducing the current in the field windings, the speed increases. Conversely, if the field is strengthened by increasing the field current, the speed decreases.

EXAMPLE
A 460-V direct current shunt motor, running on load, has an armature resistance of 0.12 Ω.

Calculate:

1. The value of the back e.m.f. when the current in the armature is 150 A.
2. The value of the armature current when the back e.m.f. is 452 V.

(CGLI 1964–5)

For a motor: $E_b = V - I_a r_a$

1. $E_b = 460 - (150 \times 0.12)$
   $= 460 - 18$ $= 442$ V

2. $I_a r_a = V - E_b$, therefore $I_a = \dfrac{V - E_b}{r_a}$

Therefore, $\quad I_a = \dfrac{460 - 452}{0.12} = \dfrac{8}{0.12} = 66.7$ A

Torque or turning effort

The torque produced by a single conductor is the product of the mechanical force tending to move the conductor about the motor shaft and the distance of the conductor from the centre of the shaft. The total torque of a motor is the sum of the separate torques of each conductor. The torque of a motor is proportional to the total flux and to the current in the armature, i.e. torque $\propto \phi I_a$.

Types of direct current motor

Direct current motors, like direct current generators, are named according to the field arrangement. There are three common types: series-wound, shunt-wound, and compound-wound.

Series-wound motor

When the motor is running, the supply current passes through both armature and field in series (see Fig. 11.9 (a)). The series winding is normally of relatively few turns of suitably large-gauge wire. The curves of torque and speed against current are shown in Fig. 11.10 (a). At low speeds the current is heavy and the turning effort is proportionately great. Thus, when starting, load, torque and current are both large. At no-load the speed tends to increase beyond the safe speed of the motor. From this it will be seen that a series motor must not be run unloaded. It is most suitable for such

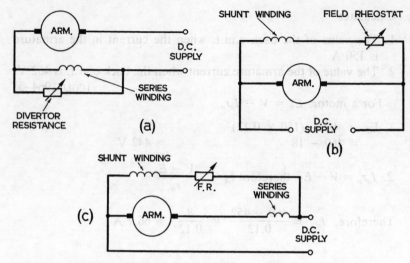

Fig. 11.9   Direct current motors: (a) Series-wound; (b) Shunt-wound; (c) Compound-wound

purposes as traction, where the motor is permanently mechanically connected to the load.

Shunt-wound motor (Fig. 11.9 (b))

The current through the field winding is approximately constant in value, from no-load to full load. Hence the speed will be fairly constant throughout the range of load. The torque is proportional to the current, is weak at low loads, and increases with increasing load (Fig. 11.9 (b)).

This motor is used for most general purposes owing to the constancy of speed. It has a good starting torque, and is used to drive machinery requiring a fairly constant speed at all loads up to full load.

Compound-wound motor

This motor has both shunt and series field windings (Fig. 11.9 c). The effect of the series winding is relatively weak compared with the shunt winding.

The series winding may be connected so as to assist the shunt winding ('cumulative compound') or to oppose the shunt winding ('differential compound'). The cumulative compound machine is

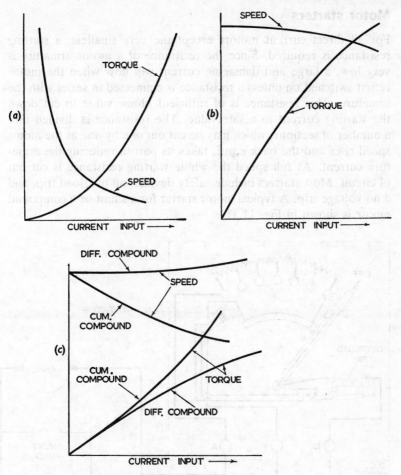

Fig. 11.10 Characteristic curves of direct current motor

used in cases where there are sudden heavy overloads as in rolling mills. The motor is fitted with a heavy flywheel so that when a sudden heavy load comes on the machine, the speed of the motor decreases and the heavy power demand is supplied by the flywheel in slowing up, reducing the sudden power demand on the mains.

The differential compound motor will give almost constant speed at all normal loads. As the load increases and the speed of the motor as a shunt motor tends to drop, the series winding weakens the field, tending to increase the speed, thus giving a steady speed throughout the range of load. Figure 11.10 (c) shows typical torque and speed curves for this motor. This type of motor is rarely used.

Motor starters

For all direct current motors except the very smallest, a starting resistance is required. Since the resistance of a motor armature is very low, a large and damaging current will flow when the motor is first switched on unless a resistance is connected in series with the armature. This resistance is of sufficient ohmic value to cut down the starting current to a safe value. The resistance is divided into a number of sections, which may be cut out one by one as the motor speed rises and the back e.m.f. takes its part in reducing the armature current. At full speed the whole starting resistance is cut out of circuit. Most starters include safety devices, an overload trip, and a no-voltage trip. A typical motor starter for a shunt or a compound motor is shown in Fig. 11.11.

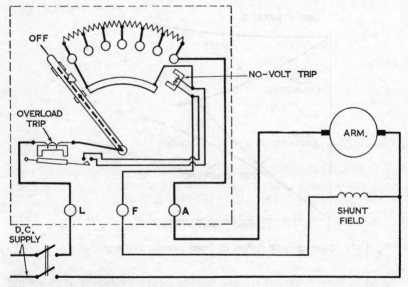

Fig. 11.11   Motor starter for direct shunt motor

No-voltage trip coil

This trip coil is a small electromagnet connected in series with the motor shunt winding. The coil is energized when the starter handle is brought up to the first resistance stud, and remains energized unless the supply fails. When the starter handle is at the end of its travel with all the starting resistances cut out of the armature circuit, the electromagnet holds firmly a small soft iron armature which is pivoted loosely to the starter arm. In the event of the supply failing by being purposely switched off or otherwise, the electromagnet is de-energized and the starter arm returns to its original position by the action of a spring. This cuts off the current to the motor, and prevents the motor restarting automatically.

Overload trip coil

This is a small electromagnet consisting of a coil of few turns of heavy wire connected in series with the armature winding. If the current passing through it is too great, the electromagnet attracts to its poles a soft iron adjustable armature pivoted at one end. This movement of the soft iron armature is arranged to short-circuit the no-voltage trip, which in turn releases the starter handle, thus cutting off the supply to the motor.

Speed control

It will be seen from the formula, $E \propto \phi N$ that provided the flux is kept constant, the speed is proportional to the back e.m.f., and therefore approximately proportional to the applied voltage V.

Series control

A series resistance connected in the armature circuit of the motor will cut down the voltage applied to the armature, and thus reduce the speed of the motor. This method is not economic and is seldom used for shunt and compound motors.

Field control

If the supply voltage is kept constant, the flux and the speed vary inversely with each other. Thus variation of speed is obtainable by altering the value of the current in the shunt winding of either a compound motor or of a shunt motor. A field regulating resistance is shown connected in Fig. 11.9 (b) and (c). Cutting out the resistance increases the field current, increases the flux, and thus reduces the speed. Weakening the field by cutting in the resistance increases the speed.

In the series motor, the effect of weakening the field is obtained by connecting a 'diverting' resistance across the field windings (Fig. 11.9 (a)).

To reverse the direction of rotation of a direct current motor

The connection of *either* the field winding *or* of the armature must be reversed.

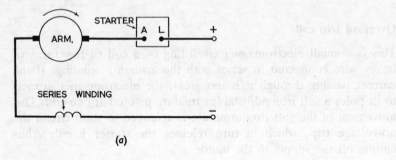

(a)

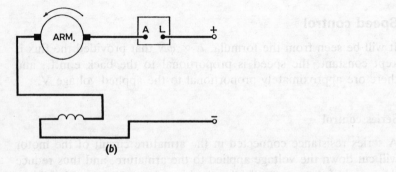

(b)

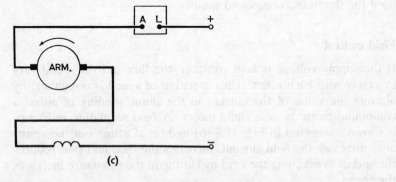

(c)

Fig. 11.12 (a) to (c) Reversing direction of rotation of direct current series motor

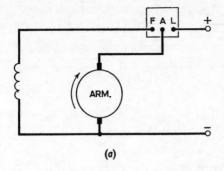

(a)

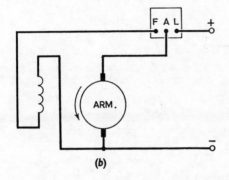

(b)

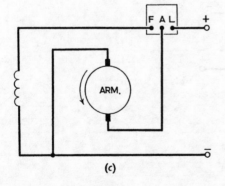

(c)

Fig. 11.13   (a) to (c) Reversing direction of rotation of direct current shunt
motor

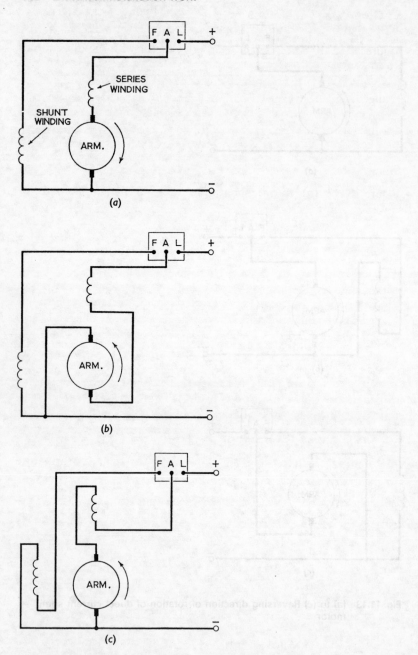

Fig. 11.14 (a) to (c) Reversing direction of rotation of direct current compound-wound motor

Series motor

Figure 11.12 (*a*) is a diagram of a series motor with the direction of rotation assumed to be clockwise.

Figure 11.12 (*b*) shows the connections to the field winding interchanged, giving reversed rotation. Figure 11.12 (*c*) shows the alternative method when the armature connections at the brushes are interchanged.

Shunt motor

Figure 11.13 (*a*) to (*c*) shows the two alternative methods of reversing the rotation of a shunt motor.

Compound-wound motor

The easiest way to reverse the direction of rotation of the compound-wound motor is to reverse the connections to the armature, Figures 11.14 (*a*) and (*b*). The reversal of the field connections involves both shunt and series windings, as shown in Fig. 11.14 (*c*).

Interpole motor

Interpoles are small poles fixed between the main poles of a motor or generator, whose function is to assist in giving sparkless commutation at all loads. The windings of these poles are connected in series with the armature. In all the above instructions the interpole windings are to be considered as part of the armature circuit. Thus the interpole connections are only reversed when the armature connections are reversed.

12 Alternating current machines

The alternator

Alternating current is used to a much greater extent than direct current. If an open-ended loop or coil of wire is rotated between the poles of an electromagnet (Fig. 12.1), an e.m.f. is generated in the loop. The value of the e.m.f. varies both in magnitude and direction according to the instantaneous position of the loop. In one revolution of the loop through 360 electrical degrees, assuming that the strength of the field is uniform, the form of the e.m.f. wave is of the shape shown in the graph (Fig. 12.2). Ideally the shape is that of a sine wave.

If slip-rings are fixed to the free ends of the loop, and sliding connections arranged to bear upon them, the alternating e.m.f. will produce an alternating current in a closed external circuit. The current will vary in a similar way to the e.m.f.

The complete change of e.m.f. or current from zero to positive maximum back to zero, and through negative maximum to zero, is called a 'cycle'. The standard frequency f of public supply is 50 hertz (Hz) or cycles per second.

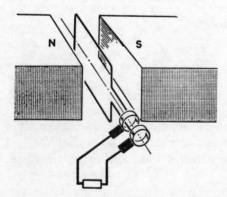

Fig. 12.1 Illustrating generation of alternating current

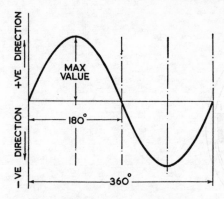

Fig. 12.2   Alternating current waveform

The practical form of the above machine is known as an alternator.

The simple alternator comprises an electromagnet with one or more pairs of poles and an armature winding wound upon a laminated core and connected to two or more slip-rings. The armature may be wound to give 1-phase, 2-phase, or 3-phase supply.

The exciter

A direct current supply is necessary to excite the electromagnets, and this is most commonly provided by a shunt-wound or compound-wound direct current generator, with field regulator, mounted on an extension of the alternator shaft.

Types of alternator

The stationary-field revolving-armature type is commonly used in small sizes and for the lower voltages. The direct current exciting current is supplied to the field windings by fixed connections, and the alternating current is delivered from the slip-rings (Fig. 12.3). The mechanical construction of the revolving armature alternator is very similar to that of the direct current generator, except that there is no commutator. The machine is usually of the open type, for convenience in attention.

The stationary-armature revolving-field type is invariably used in the generation of high voltages. The main reason for this type is the

REVOLVING   ARMATURE

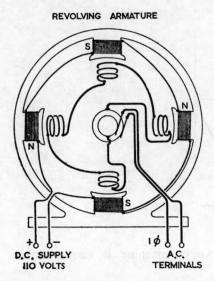

D.C. SUPPLY
110 VOLTS

1φ
A.C. TERMINALS

Fig. 12.3 Stationary-field revolving-armature type alternator

REVOLVING   FIELD

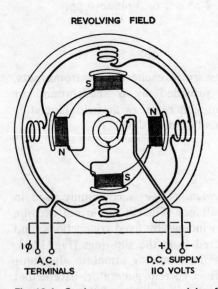

1φ
A.C. TERMINALS

D.C. SUPPLY
110 VOLTS

Fig. 12.4 Stationary-armature revolving-field type alternator

difficulty of using sliding contacts – brushes bearing on slip-rings – at high voltages. With a stationary armature, the power from the generator is delivered through copper-to-copper connections firmly bolted together. The revolving fields are supplied with direct current, normally at 110 V, through a pair of slip-rings (Fig. 12.4). The mechanical construction of the high-speed turbo-alternator with revolving field is necessarily different from the slow-speed type. The rotor is long and of small diameter. The rotor core may be built of laminated stampings bolted together, or machined from a solid steel casting. The stationary armature core is made up of stampings bolted together. The machine is normally totally enclosed, and may be cooled by forced draught.

Frequency

The frequency of the current is given by the formula:

$$f = pn$$

where f is the frequency in Hz

p is the number of pairs of poles

n is the speed of the armature in revolutions per second (rev/s)

The following table shows a few rotor speeds for the standard frequency of 50 Hz

2-pole: $50 = 1 \times n$, $n = 50$ rev/s, or 3000 rev/min
4-pole: $50 = 2 \times n$, $n = 25$ rev/s, or 1500 rev/min
12-pole: $50 = 6 \times n$, $n = 8\frac{1}{3}$ rev/s, or 500 rev/min

The earlier alternators were driven by slow-speed reciprocating steam engines, and thus needed a large number of poles. The modern high-voltage alternators in large generating stations are driven by high-speed steam turbines running at 25 or 50 rev/s, and are consequently 4-pole or 2-pole respectively.

E.M.F. formula

The e.m.f. per phase is given by the formula,

$E = K\phi Znp$ volts

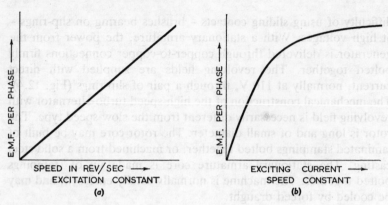

Fig. 12.5   (a) to (b) Characteristic curves of alternating current generator

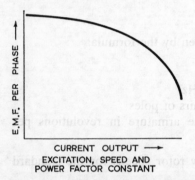

Fig. 12.6   Load characteristic of a.c. generator

where: K is a constant dependent upon various factors in the design of the machine,

ϕ is the total flux per pole, in webers

n is the speed in rev/s

p is the number of pairs of poles

Z is the number of conductors in series per phase

From the formula it will be seen that the e.m.f. is directly proportional to the speed of the rotor (armature or field as the case may be) (Fig. 12.5 (a)).

The e.m.f. is also proportional to the flux per pole; in other words, the e.m.f. will vary with the exciting current (Fig. 12.5 (b)).

Figure 12.6 shows a typical load characteristic, the variation of e.m.f. with current loading at constant excitation and at constant power factor.

Alternating current motors

Single-phase motors

A single-phase motor with its starting equipment is more costly, particularly in the larger sizes, than the equivalent 3-phase motor and equipment. In addition, supply companies severely limit the size of single-phase motors that they will allow to be connected to their mains. For these reasons the use of single-phase motors is restricted in general to very small motors for domestic and similar uses, and up to a limit of, say, 8 kW for industrial purposes. Two main types of motor are available: the commutator motor and the induction motor. The common types of commutator motor produce a torque which is high at starting and at low speeds, but with light load the speed rapidly increases to what may be a dangerous value. The induction-type motor, on the other hand, has a speed-torque characteristic similar to that of the direct current shunt motor and the speed is fairly constant from no-load to full load.

Single-phase series motor

If the direct current supply to a direct current series motor is reversed, the motor will continue to run in the same direction. Thus a series motor is usable on single-phase alternating current. The 'universal' motor as used in such articles as vacuum cleaners will run with either direct current or alternating current. For satisfactory use on alternating current the series motor must have its magnetic circuit laminated throughout – poles, yoke, and armature core. This motor, which has, of course, a series speed-torque characteristic,

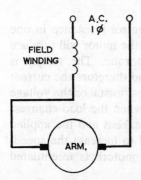

Fig. 12.7   Single-phase series motor

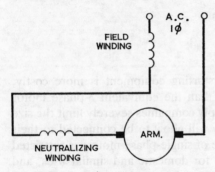

Fig. 12.8 Neutralized series motor

is used in such articles as gramophone turn-tables, vacuum cleaners, etc., where the load is approximately constant and where the motor does not run unloaded. Figure 12.7 is a diagram of the simple single-phase series motor. To improve commutation a further field winding, the neutralizing winding, is added (Fig. 12.8). The neutralized series motor is used in larger sizes for single-phase traction work and for other industrial purposes where a motor with a series speed-torque characteristic is required.

Starting

Direct-on starting is usual for small motors. For larger motors a simple resistance starter in one line is used in order to cut down the starting current.

Speed control

This is obtainable by the use of a series control resistance in one supply line. The reduction of voltage across the motor will produce a reduction in the speed for any given torque. The torque is proportional to the square of the current, and therefore the current varies as the load changes. With a given series resistance the voltage applied to the motor terminals will change when the load changes. If the load and the current fall, the field weakens and the applied voltage rises; both these circumstances tend to increase the speed. This behaviour, which is typical of all series motors, is accentuated by the presence of a series resistance.

Reversing the direction of rotation

For this purpose the leads to *either* the field winding *or* the armature should be reversed.

Single-phase repulsion motor

The single-phase repulsion motor comprises a single-phase winding wound on a stator, and an ordinary commutator type wound armature with sets of brushes which are short-circuited.

When a single-phase supply is applied to the stator winding, the machine acts as a transformer and induced currents flow in the rotor and through the short-circuiting connection. If the brushes were set in line with the stator field (perpendicular in the diagram, which relates to a 2-pole machine) (Fig. 12.9), there would be no turning moment or torque on the rotor, as the field set up by the stator and that set up by the rotor would be in opposition. On the other hand, if the brushes were set at right-angles to the stator field (horizontal in the diagram), no current would flow in the rotor since the brushes are in the no-voltage position, and so no torque would be created. For satisfactory operation, therefore, the brushes are placed between these positions as in the diagram, and the resulting reaction between the fields creates a torque which causes the rotor to revolve.

The repulsion motor has a series characteristic, but is most suitable for a fixed speed drive where good starting qualities are essen-

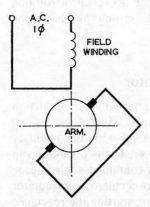

Fig. 12.9 Single-phase repulsion motor

tial. The repulsion motor is not limited to small sizes as is the series motor but is also manufactured in medium sizes.

Reversing the direction of rotation

By moving the brushes round the commutator, both the starting torque and the direction of rotation may be controlled.

Starting

In the small sizes direct-on starting is employed, but in the medium sizes a simple series resistance starter in the line is necessary to cut down the starting current.

Single-phase repulsion-start induction motor

This motor combines the good starting properties of the repulsion motor with the approximately constant speed of the induction motor. It starts as a repulsion motor, but at about 70 per cent of synchronous speed a centrifugal mechanism brings the short-circuiting ring to bear against the whole of the commutator segments, and also moves the brushes away from the face-plate commutator, thus converting the rotor into a squirrel-cage rotor (see Fig. 12.10). The machine then runs normally as a squirrel-cage induction motor.

The motor is made in small sizes only, and may be started direct-on or by means of a series resistance starter. Reversal of rotation is obtainable by alteration of the brush setting as shown on Fig. 12.11.

Single-phase repulsion-induction motor

This is an improvement on the foregoing repulsion-start induction motor, in that the same effects are obtained without the use of brush-lifting mechanism. The stator is wound with the ordinary repulsion motor winding. The rotor, however, has two windings in common slots. The inner winding is a self-contained squirrel-cage winding, and the outer winding is a short-circuited, commutator type repulsion winding. At low speeds during starting the reactance of the squirrel-cage winding is high, and this winding has little effect upon starting. The motor starts, therefore, as a repulsion motor

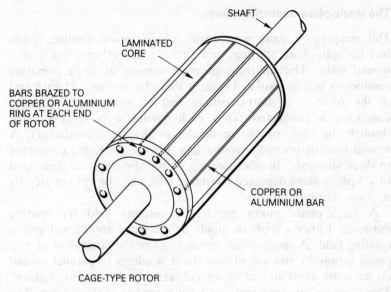

Fig. 12.10

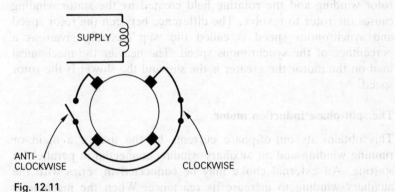

Fig. 12.11

with good starting torque. At higher speeds the reactance of the squirrel-cage winding is proportionally reduced, and its torque, which is that of an induction motor, is superimposed upon the torque of the repulsion motor, which remains in operation. The overall result is to give good starting qualities, and a fairly constant speed from no-load to full load, and a reasonably good power factor. The speed is near to synchronism throughout the working range and the efficiency is very good.

Starting as before is either direct-on or by series resistance starter.

The single-phase induction motor

This comprises a stator wound with a single-phase winding, modified for split-phase starting, and either a squirrel-cage rotor or a wound rotor. The squirrel-cage rotor consists of single armature conductors laid in insulated slots in the rotor surface. At each end of the rotor is a short-circuiting ring to which each armature conductor is connected. Thus each conductor is short-circuited through the rings by the remainder of the rotor conductors. A wound rotor is normally wound as a 3-phase rotor and connected to three slip-rings. Brushes bearing on the slip-rings are connected to a 3-phase star-connected resistance which can be cut out step by step for starting purposes.

A single-phase motor requires a rotating field for starting purposes. Either a 2-phase supply or a 3-phase supply will give a rotating field. A single-phase motor is given the equivalent of a 2-phase supply by the use of two stator windings in parallel wound on the stator at 90° to each other and carrying currents out of phase. This '2-phase' arrangement can be obtained in various ways. The reaction between the field created by the induced currents in the rotor winding and the rotating field created by the stator winding causes the rotor to revolve. The difference between the rotor speed and synchronous speed is called the 'slip,' usually given as a percentage of the synchronous speed. The heavier the mechanical load on the motor the greater is the slip and the slower is the rotor speed.

The split-phase induction motor

This obtains its out-of-phase currents by the use of a main or running winding and an auxiliary winding connected in parallel for starting. An external choke may be connected in series with the auxiliary winding to increase its reactance. When the machine is running the main winding only is left in circuit, the auxiliary winding and its external choke being cut out. This can be done by means of a 3-position starting switch, or automatically by a centrifugally operated mechanism on the rotor shaft (Fig. 12.12).

The split-phase induction motor has a shunt characteristic and runs at fairly constant speed at all loads within its working range without possibility of speed control. It will not start against load, and is most useful for drives which require a reasonably constant speed, such as machine tools, circular saws, etc. A reduction of voltage has little effect on the speed but causes a reduction of torque. Reversal of rotation is obtained by reversing either the

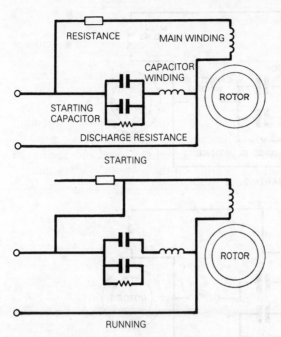

Fig. 12.12   Split-phase induction motor

starting winding or the running winding. The motor must be brought to a standstill for this purpose.

Starting

Small motors may be started by the direct-on method. Larger squirrel-cage motors may be started by a step-by-step resistance starter in series with the running winding. Wound rotor motors are started by cutting out resistance step by step in the rotor external resistance until the rotor is running at normal speed.

The single-phase capacitor type induction motor

This is generally similar to the split-phase type above, except that the starting winding contains a capacitor instead of an inductance. By the use of capacitance, a greater phase displacement between the currents in the starting and running windings is obtained, giving greater starting torque with smaller line current. The capacitor may be used for power-factor correction when the motor is on load. Figure 12.13 shows one of the possible methods of connection. The

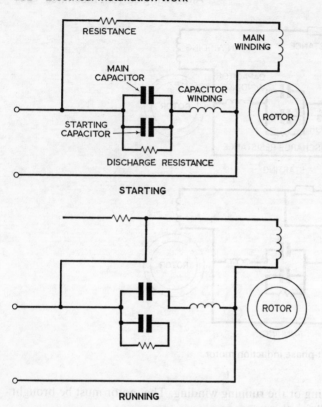

Fig. 12.13 Single-phase capacitor type induction motor

two capacitors are used in parallel during starting. In the running position the main capacitor is left connected to provide power-factor correction, and the starting capacitor is disconnected and discharged through the discharge resistance.

Starting is by direct-on switch or by starting resistance in the line for squirrel-cage machines, or by rotor external resistance for wound rotor machines.

Reversal of rotation is obtained by reversing the connections of either the main winding or the starting winding.

The single-phase synchronous motor

Since the inception of the British Grid with its accurate frequency control, small synchronous motors have come into wide use as electric clock drives and for similar purposes. One form of clock motor

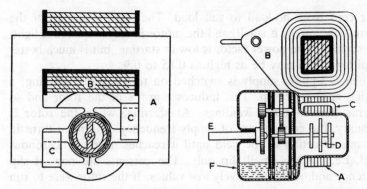

Fig. 12.14   Single-phase synchronous clock motor (Everett Edgcumbe & Co Ltd)

consists of a laminated iron circuit on which is mounted a magnetizing coil excited by single-phase alternating current. The pole pieces are each split into two parts, one part being 'shaded' by embracing it with a closed band of copper. The pole flux is thus split into two parts out of phase with each other. The rotor is simply a slotted steel bar or a number of discs (Fig. 12.14). The 2-phase flux creates a rotating field, causing the rotor to revolve. When the motor reaches a speed near to synchronism, the rotor locks into synchronous speed. This particular type of motor is self-starting.

Three-phase motors

Where a 3-phase supply is available, 3-phase motors are preferred to single-phase motors in all but the smallest sizes. These motors are more efficient, have a higher power factor, have better starting properties, and are cheaper than the corresponding single-phase motors. Supply authorities require 3-phase motors to be used where possible, to prevent unbalanced loads on the mains.

The squirrel-cage three-phase induction motor

This motor has a wound 3-phase stator, and a short-circuited cage rotor. There are no slip-rings and thus no moving connections. The motor is robust, simple, and cheap. It is essentially a single speed motor, with a 'shunt' speed-load characteristic. Very large motors are made which have a speed drop or slip of about 1.5 per cent from no-load to full-load. Smaller motors have a maximum slip of about

4 per cent from no-load to full load. The starting torque of the squirrel-cage motor is small, and the motor is only suitable for light-duty starting. The power factor is low at starting, but is much better at full load, and may be as high as 0.85 to 0.9.

When a 3-phase supply is switched on to the stator winding, a rotating field is set up. This induces e.m.f.'s in the rotor and so currents flow in these windings. At starting, when the rotor is stationary, the currents are of supply frequency. The rotor turns in the same direction as the field until it reaches nearly synchronous speed, i.e., with a small slip only. The currents are now of slip frequency and of comparatively low values. If the rotor were to run at synchronous speed, no currents would flow in the rotor, and there would be no torque to turn the rotor.

When the motor is mechanically loaded, the rotor slows down slightly, the frequency and value of the rotor currents increase, and more power is taken by the stator from the mains.

Starting

If the motor is started by direct-on switch, the current is large and of a low power factor. For all but the smallest motors it is necesary to reduce the starting current. With the ordinary squirrel-cage motor this can only be done by reducing the voltage supplied to the stator. The torque is proportional to the square of the voltage and so is small at the reduced voltage. Line resistance starting is sometimes used but it is inefficient.

Direct-on-line starting

For small machines direct-on-line starting is allowable. Figure 12.15 shows the basic connections for one type, When the START button is pushed the magnetic operating coil is energized from two of the line conductors, and four sets of contacts close, providing current to the motor, and also short-circuiting the START button. The operating coil holds the contacts in place while the motor is running and thus the motor is stopped by pressing the STOP button. This action breaks the operating circuit, de-energizes the operating coil, breaking the contacts and stopping the motor.

The overload coils, which are either thermal or magnetic, are so arranged that in times of undue overload they will open a switch in the operating circuit and stop the motor.

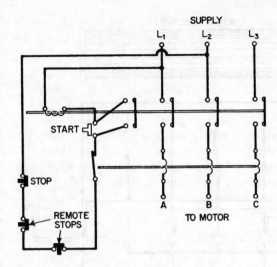

Fig. 12.15   Direct-on-line starter

Star-delta starting

This is a cheap and simple method of squirrel-cage motor starting. The ends of each phase of the stator winding, six in all, are brought out to a change-over switch. In the starting position the windings are connected in star so that each phase receives $1/\sqrt{3}$ of the normal supply voltage. The current and torque are reduced in the square of the voltage ratio, that is, to one-third of the direct-on value. When the motor is running at or near its normal speed, the switch is changed over so as to connect the windings in delta or mesh, when each phase receives full mains voltage. A diagram of connections is given in Fig. 12.16.

Auto-transformer starting

In this method of starting the voltage reduction at starting is obtained by means of a 3-phase auto-transformer, by connecting the stator to suitable tappings on the secondary side of the transformer by means of a tapping switch. Although the transformer may have a number of tappings, normally one reduced voltage tapping only is used. Figure 12.17 shows the connections of a suitable tapping switch.

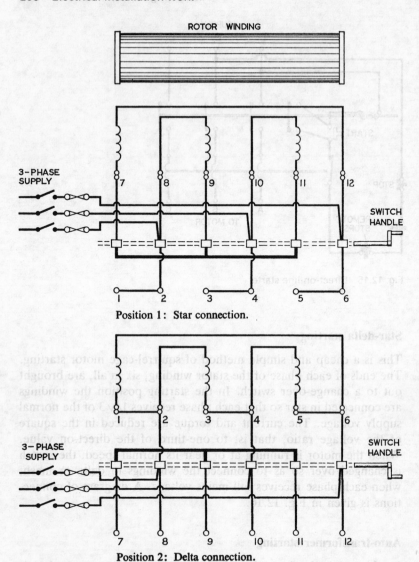

Position 1: Star connection.

Position 2: Delta connection.

Fig. 12.16 Star-delta starting of 3-phase squirrel-cage motor

Reversal of rotation

Any 3-phase motor may have its direction of rotation reversed by interchanging any two of the three line connections to the motor.

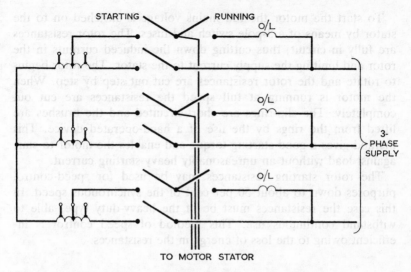

Fig. 12.17   Auto-transformer starting of 3-phase induction motor

The wound rotor type three-phase induction motor

The stator winding is wound 3-phase as in the case of the squirrel-cage motor. The rotor usually consists of a slotted armature with a 3-phase star-connected winding. Mesh connection can equally well be used, although it is less common. The ends of the phases are brought out to three slip-rings. Brushes bearing on the rings connect the phases to a tapped 3-phase resistance. Figure 12.18 shows the connections of the rotor starter.

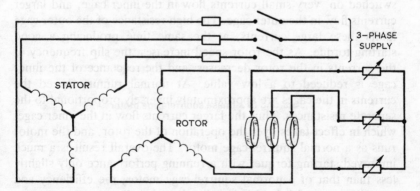

Fig. 12.18   Rotor starting of 3-phase induction motor

To start the motor the full mains voltage is switched on to the stator by means of a 3-pole switch and fuses. The rotor resistances are fully in circuit, thus cutting down the induced currents in the rotor and limiting the supply current to the stator. The rotor begins to rotate and the rotor resistances are cut out step by step. When the motor is running at full speed the resistances are cut out completely. The slip-rings are short-circuited and the brushes are lifted from the rings by the use of a hand-operated device. This method gives a good starting torque and enables the motor to start against load without an unreasonably heavy starting current.

The rotor starting resistances may be used for speed-control purposes down to about 50 per cent of the synchronous speed. In this case the resistances must be of the heavy-duty type, able to withstand continuous use. This method of speed control is inefficient owing to the loss of energy in the resistances.

Double-cage squirrel-cage three-phase induction motor

This motor, commonly named the high-torque motor, has been developed to retain the robust qualities and the simplicity of the ordinary squirrel-cage motor, and to approach the performance of the slip-ring motor in starting ability. The double-cage motor comprises a normal 3-phase stator and a rotor with two separate squirrel-cages. The inner cage winding is of low resistance and of high reactance at low rotor speeds, that is, at high slip frequency. It is wound in deep slots in the rotor surface. The outer cage winding is wound in shallow slots in the rotor surface, and is of high resistance and low reactance. When the supply to the stator is switched on, very small currents flow in the inner cage, and larger currents flow in the outer cage. The high resistance of the outer cage limits these large currents, at the same time producing a good starting torque. As the rotor speed increases, the slip frequency of the currents in the rotor decreases, and the reactance of the inner cage is reduced to a low value. At normal running speed the currents in the cages are approximately inversely proportional to the separate resistances. Thus the larger currents flow in the inner cage, which in effect takes over the operation of the rotor, and the motor runs as a normal squirrel-cage motor. The overall result is a much improved starting torque, with a running performance only slightly less than that of a normal squirrel-cage motor, the efficiency and power factor at full load being slightly reduced.

The variable speed commutator type three-phase induction motor

The machine here described works on the Schrage system of speed control. Briefly, it consists of an ordinary wound slip-ring induction motor with primary and secondary windings reversed in position. The primary winding is fixed into slots on the rotor and connected direct to the line starting switch through slip-rings, while the secondary winding is fixed in slots on the stator. In addition, there is a third (regulating) winding in the same rotor slots as the primary winding, and this winding is connected to commutator segments on the rotor shaft. There are two brush rockers, each holding three separate brush arms. From the three brush arms on one rocker connections are made to one end of each of the secondary windings, and from the brush arms on the other rocker connections are made to the other ends of the secondary windings (see Fig. 12.19). The brush arms are movable round the commutator and are geared together so as to move in opposite directions. The respective pairs of brushes may be 'in line', that is, on the same commutator segment, or may be moved in either direction to include the desired

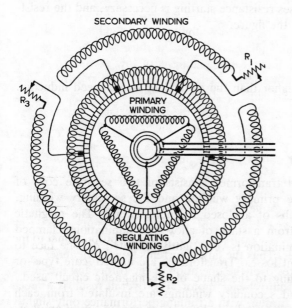

Fig. 12.19   Variable speed 3-phase induction motor (Metropolitan-Vickers Electrical Co Ltd)

number of segments. Thus the e.m.f.'s generated in the separate coils of the regulating winding may be tapped off in varying values and injected into the secondary windings.

When the brushes are in line, the secondary windings are short-circuited on themselves as in the induction motor proper, and the motor will run as an ordinary wound rotor induction motor at a speed just below synchronous speed, that is, with a small value of slip. When the brushes are in such a position that the e.m.f. from the regulating winding opposes the current in the secondary winding, the motor slows down so that the increased e.m.f. due to the increased slip balances the injected e.m.f. If the brushes are moved over in the opposite direction, the motor will speed up above synchronous speed for similar reasons.

The normal motor of this type is arranged to have a speed range of 2:1 or 3:1, although motors with speed ranges up to 15:1 are manufactured.

These motors have a shunt characteristic, the speed in any one rocker position being only slightly affected by increase of load.

Starting is simply effected in the normal motor by placing the brushes in line and switching on through the usual 3-pole starting switch. In some cases resistance starting is necessary, and the resistances are shown in the figure.

Power factor

This is generally higher than the equivalent single-speed induction motor.

The transformer

The double-wound transformer consists of two separate coils of insulated wire, the primary winding and the secondary winding, wound on the limbs of a closed magnetic circuit. The magnetic circuit is formed from a stack of alloy iron laminations clamped together. Each lamination is very tightly insulated on one face to reduce eddy-current losses. Transformers are named 'core type' or 'shell type' according to the shape of the magnetic circuit used.

The primary and secondary windings are insulated from each other, and may be arranged side by side, or one on top of another. Figure 12.20 illustrates core and shell type single-phase transformers.

An alternating voltage applied to the primary winding will

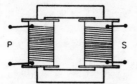 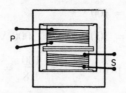

Fig. 12.20   Core and shell type transformers; single phase

produce an alternating magnetic flux in the iron, which will in turn induce an alternating e.m.f. in the secondary winding. If the losses in the transformer are ignored, the ratio of the two voltages will equal the ratio of the turns in the two windings. Since the flux is common to the two windings, the 'volts per turn' will be the same in each winding.

When a load is connected across the secondary terminals, a current will be taken from the winding and a reverse magnetic flux is created which tends to reduce the original value of the flux. To redress the balance, current must flow from the supply to the primary winding to restore the flux.

Since the 'volts per turn' are the same in each winding, and the power in each winding must be equal, then we have in the form of a formula:

$$\frac{V_1}{V_2} = \frac{N_1}{N_2} = \frac{I_2}{I_1},$$

where: V_1 and V_2 represent the primary and secondary voltages respectively

N_1 and N_2 the primary and secondary numbers of turns

I_1 and I_2 the primary and secondary currents

A 'step-down' transformer is one in which the primary turns are the greater number, and a 'step-up' transformer one in which the secondary turns are the greater number.

Transformers are always rated in kVA because this is the maximum demand that can be imposed upon the transformer, irrespective of the power factor of the load connected to the secondary turns of the transformer.

EXAMPLE
The 'volts per turn' of a simple transformer is 1.8. Calculate the respective number of turns in each winding of a single-phase transformer with a step-up ratio of 1 to 30, and a primary voltage of 100. Ignore transformer losses.

$$V_1 = 100$$

therefore $V_2 = 100 \times 30 = 3000$ V

No. of turns in primary $= \dfrac{100}{1.8} = 56$ turns

No. of turns in secondary $= \dfrac{3000}{1.8} = 1667$ turns

EXAMPLE

A single-phase step-down transformer has a transformation ratio of 15 to 1. The primary winding supplied at 3000 V has 1650 turns. Ignoring losses, calculate the secondary voltage, and the number of secondary turns.

Secondary voltage $= \dfrac{3000}{15} = 200$ volts

No. of secondary turns $= \dfrac{1650}{15} = 110$ turns

Transformer efficiency

The previous arguments assume no losses in the transformer. However, there are losses, and these are of two kinds. Those losses due to the currents in the primary and secondary windings passing through the resistances of the windings, can be lumped together as RI^2 losses or 'copper losses'. These losses are in proportion to the square of the load current.

The other losses are those due to the magnetization of the transformer core. They comprise 'eddy-current' and 'hysteresis' losses. For practical purposes they are lumped together as 'iron losses', and are considered to be of constant value irrespective of the load on the transformer. The losses can be found by two tests, the 'open-circuit test' and the 'short-circuit test'.

Open-circuit test

In this test, readings are taken at normal supply voltage with the secondary terminals on open circuit. The wattmeter reading on the primary side is taken as the measure of the total iron losses in the transformer.

Short-circuit test

The secondary terminals are short-circuited through an ammeter. The primary winding is supplied with a considerably reduced voltage of such a value that the current in the short-circuiting ammeter reads full-load current. The wattmeter reading on the primary side is taken as the measure of the copper losses in the transformer. *The efficiency of a transformer* is calculated from the formula:

$$\text{Efficiency at full load} = \frac{(VA \times p.f.)}{(VA \times \text{p.f.}) + \text{copper losses} + \text{iron losses}}$$

EXAMPLE

A 20-kVA transformer was found to have 600 W iron losses, and 700 W copper losses, when supplying full load at 0.8 power factor.

Calculate the efficiency of the transformer at 0.8 power factor:

1. on full load,
2. on half full load.

Copper losses on full load $= 700$ W
Copper losses on half full load $= 700 \times \frac{1}{2}^2 = 175$ W
Iron losses, constant $= 600$ W

1. Efficiency on full load $= \dfrac{(20\,000 \times 0.8)}{(20\,000 \times 0.8) + 700 + 600} \times 100\%$

   $= \dfrac{16\,000}{17\,300} \times 100\%$

   $= 92.5\%$

2. Efficiency on $\frac{1}{2}$ full load $= \dfrac{(10\,000 \times 0.8)}{(10\,000 \times 0.8) + 175 + 600}$

   $= \dfrac{8000}{8775} \times 100\%$

   $= 91.2\%$

Cooling of transformers

The losses in a transformer are produced as heat. To prevent dangerous rises of temperature in transformers, some method of artificial cooling is usually necessary. In the very smallest transformers the heat is radiated away from the transformer case with no difficulty. The larger, three-phase transformers, are mainly equipped with oil cooling. The transformer is immersed in a suitable steel tank, the tank being constructed with fins or tubes to facilitate the circulation of the warmed oil from the heated coils of the transformer to the cooler walls of the tank. Other measures, such as air-blasting are used for the largest transformers. Where oil cooling is used, special precautions must be taken for fire prevention and for fire extinction.

The auto-transformer

This transformer has one winding only, the primary winding. The secondary voltage is obtained by tapping the winding at suitable points. Figure 12.21 shows a 2-to-1 auto-transformer and the current distribution in its windings. It is used because of its cheapness, but has limited uses. A common use is in the auto-transformer starting of induction motors. A disadvantage of this transformer is the fact that in the event of a break or disconnection in the common section of the winding, the full primary voltage will be present at the lower-voltage secondary terminals.

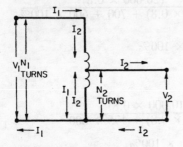

Fig. 12.21   Auto-transformer

The current transformer

If a coil of insulated wire with an ammeter connected across its ends is fixed around a conductor carrying alternating current, the small e.m.f. generated in the coil will cause a current to flow in the ammeter. This fact is used to measure the current in high voltage lines or to operate trip circuits at safe potentials. The reading on the ammeter will vary according to the current variation in the high voltage line – the primary winding of the transformer. These transformers are rated according to their current ratios, e.g. 200 to 5,100 to 5, and so on. The secondary winding is always rated at 5 A.

This transformer must not be operated with the secondary winding on open circuit. If it is desired to remove an ammeter for repair or replacement, the secondary terminals must be short-circuited before the ammeter is disconnected. The short-circuit must be kept in place until the ammeter is again connected in place. If the secondary terminals are left open for even a brief time, the voltage may rise to a value sufficiently high to damage the windings. Figure 12.22 illustrates a current transformer.

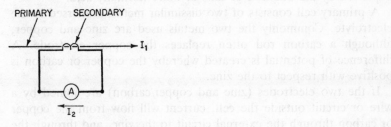

Fig. 12.22   Current transformer

13 Primary cells

Primary cells

One way in which electricity may be generated is by the chemical action between metals and an electrolyte. An electrolyte is a liquid that conducts electricity, and this ability to conduct electricity is improved by the addition of salts or more normally acid.

A primary cell consists of two dissimilar metals immersed in the electrolyte. Commonly the two metals used are zinc and copper, although a carbon rod often replaces the copper electrode. A difference of potential is created whereby the copper or carbon is positive with respect to the zinc.

If the two electrodes (zinc and copper/carbon) are joined by a wire or circuit outside the cell, current will flow from the copper or carbon through the external circuit to the zinc, and through the cell from zinc to copper/carbon. In this way the chemical energy stored in the materials used in the cell is converted into electrical energy.

In some simple forms of primary cells, the chemical action takes place whether the cell is supplying current to an external circuit or not. In commercial forms of primary cells, little or no chemical action should take place when cells are not in active use, this allowing a reasonable commercial 'shelf life'.

Leclanché cell 'wet' type

There are several types of primary cell, each with its own particular advantages. A now outmoded type that was widely used in the installation of bells and telephones was Leclanché cell.

Leclanché cell, 'dry' type (Fig. 13.1)

This type of Leclanché cell is useful because of its portability, and compact form.

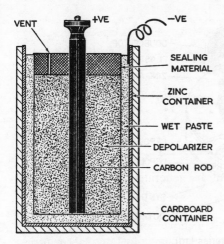

VENT +VE −VE

SEALING MATERIAL

ZINC CONTAINER

WET PASTE

DEPOLARIZER

CARBON ROD

CARDBOARD CONTAINER

Fig. 13.1 Leclanché cell, 'dry' type

The positive plate is a carbon rod, surrounded with a mixture of manganese dioxide and carbon. Around this is a wet paste of plaster of Paris, flour, and zinc chloride. The outside vessel is of zinc and forms the negative plate. The whole is surrounded by cardboard, and the top is filled in with sand or other insulating material and covered with a sealing compound. A vent is also provided.

The action is similar to that of the 'wet' cell.

The e.m.f. is about 1.6 volts when new. It quickly falls to 1.5 volts and remains at this voltage until the cell is near to exhaustion. When the e.m.f is about 1 V the cell should be discarded. The internal resistance is low, from 0.1 to 0.3 Ω.

Mercury cells

The mercury cell is manufactured particularly in miniature form as a source of power for use with hearing aids, wrist-watches, clocks, photography and the like. It is also important in terms of longer storage and long use period. In basic form the mercury cell consists of an outer sealed steel container which is the positive electrode, and a cylinder of compressed zinc powder set in the centre of the cells which is the negative electrode. The negative electrode is surrounded by a potassium hydroxide (alkaline) electrolyte and a depolarizer mix of manganese dioxide and carbon pressed in cylindrical form. The nominal voltage is about 1.35 V. After a slight

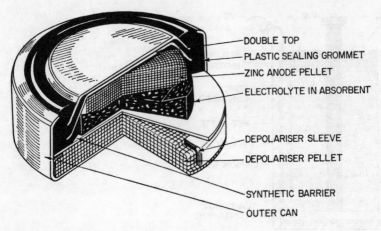

DOUBLE TOP
PLASTIC SEALING GROMMET
ZINC ANODE PELLET
ELECTROLYTE IN ABSORBENT
DEPOLARISER SLEEVE
DEPOLARISER PELLET
SYNTHETIC BARRIER
OUTER CAN

Fig. 13.2   Mercury cell (Mallory Batteries Ltd)

initial voltage drop the output voltage remains practically constant until near the end of its practical life when the voltage drops away very quickly. The life of this cell is closely predictable under defined conditions. Figure 13.2 shows the mercury cell in button form.

A recent development of the mercury cell is the 'reserve cell'. This remains completely passive until a screw is turned at the top of the cell, an action which breaks a sealed electrolyte container to activate the cell. It is then capable of providing practically the same capacity as an ordinary mercury cell of the same size. It has the advantage that it can be stored for many years in temperatures ranging from − 50 °C to + 50 °C without deterioration.

These details and the illustration of a mercury cell are given by permission of the makers, Mallory Batteries Limited.

Silver oxide cell

Following the development of the mercury cell came the silver oxide cell, both cells having very similar construction as shown in Figs 13.2 and 13.3.

The advantages of this silver oxide cell is its reduced physical size for the same output, a higher e.m.f. (1.5 V) and an excellent 'shelf life'.

The silver oxide cell consists of a depolaring silver-oxide cathode connected to the stainless steel or nickel-plated steel casing of the cell, which is the positive terminal. Powdered zinc forms the anode,

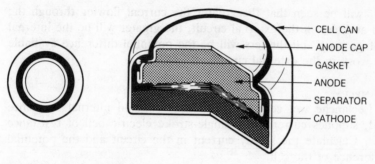

Fig. 13.3   Silver oxide cell

and as this anode has a comparitively large surface area its internal resistance is reduced. A highly alkaline electrolyte is used.

This cell has a longer service life than the mercury cell, and is widely used to power hearing aids, electronic wrist watches and scientific instruments.

Internal resistance of a cell

Every primary cell has electrical resistance, comprising the resistance of the positive and negative plates and the resistance of the electrolyte. This is called the 'internal resistance of the cell'. This resistance will vary with the type of cell and also with its dimensions. For instance, an increase in the size of the plates, or a decrease in the distance between them, will both decrease the internal resistance. This internal resistance will have an effect on the potential difference of the cell, when the latter is supplying a current.

The e.m.f., E, of a cell is the total voltage generated by the cell, and is measured with the cell on open circuit. The test is made with a high-resistance voltmeter, so that the current taken by the voltmeter is negligible.

When the cell is delivering current to an outside circuit, the current I flowing in the external circuit will necessarily flow through the cell itself. This current creates a voltage drop rI in the cell. The potential difference V of a cell is the e.m.f. less the internal voltage drop.

Let r be the internal resistance, and I the current flowing.

Then, $V = E - rI$.

It will be seen that the greater the current flowing through the cell and also in the external circuit, the greater will be the internal voltage drop and the less will be the potential difference available for work in the external circuit.

EXAMPLE
A Leclanché wet cell, with e.m.f. 1.5 V, and internal resistance 1 Ω, supplies current to a single-stroke electric bell of resistance 5 Ω. Calculate the steady current in the circuit and the potential difference of the cell.

Total resistance in circuit = $R + r = 5 + 1 = 6 \, \Omega$

$$I = \frac{\text{e.m.f.}}{\text{total resistance}}$$

Therefore, Current in circuit $= \dfrac{1.5}{6} = 0.25$ A

Potential difference of cell = $E - rI = 1.5 - (1 \times 0.25) = 1.25$ V

Check Calculation. In external circuit, $V = RI$.

Therefore, $V = 5 \times 0.25 = 1.25$ V

Methods of connecting a number of cells

It is frequently necessary to use two or more cells to get a greater output than from one cell only. There are three ways in which cells may be connected:

1. In series
2. In parallel
3. In series-parallel

Series connection

Let n cells be connected in series, with the positive terminal of one to the negative terminal of the next, as in Fig. 13.4. The free terminals of the two end cells will be the positive and negative terminals of the battery. Any two terminals which are connected together will be at the same potential or electric level. Therefore, the positive terminal of any one cell will be at a potential E volts higher than that of the preceding cell. Thus the resultant e.m.f. of the battery will be the sum of the separate e.m.f.'s. Also, since any

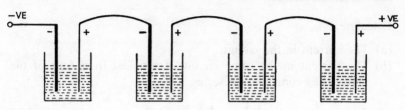

Fig. 13.4

current flowing in the circuit will pass through each cell in turn, the internal resistances must be added together:

Total e.m.f. = nE, and total internal resistance = nr.

Whether it is worth while to connect a number of cells in series in order to increase the current in the circuit, depends not only on the internal resistances but also on the external resistance of the circuit.

If the external resistance is small compared with the internal resistance, so that the resistance of the circuit is mainly due to the cells themselves, series connection will be of little avail, since there will be little increase of current in the circuit. If, however, the external resistance is large compared with the internal resistance, the current will be increased almost in proportion to the increase of e.m.f. The following examples show these points clearly.

EXAMPLE

A primary cell with e.m.f. of 1.5 V and internal resistance 0.2 Ω is connected to a circuit of 20 Ω resistance.
Calculate:

(a) The current flowing in the circuit.
(b) The current in the circuit if supplied from ten similar cells connected in series.

(a) Current = $\dfrac{\text{total e.m.f.}}{\text{total resistance}} = \dfrac{E}{R + r} = \dfrac{1.5}{20 + 0.2} = \dfrac{1.5}{20.2} = 0.074\ A$

(b) $I = \dfrac{nE}{R + nr} = \dfrac{10 \times 1.5}{20 + (10 \times 0.2)} = \dfrac{15}{22} = 0.682\ A$

Note: The current in (b) is over nine times greater than in (a).

EXAMPLE

A primary cell with e.m.f. 1.4 V, and internal resistance 1 ohm, is connected to a circuit of resistance 0.4 Ω

Calculate:

(a) The current in the circuit.
(b) The current in the same circuit, if supplied from four of the above cells connected in series.

(a) $I = \dfrac{E}{R + r} = \dfrac{1.4}{0.4 + 1} = \dfrac{1.4}{1.4} = 1\ \text{A}$

(b) $I = \dfrac{nE}{R + nr} = \dfrac{4 \times 1.4}{0.4 + (4 \times 1)} = \dfrac{5.6}{4.4} = 1.27\ \text{A}$

Note: The current with four cells in series is only one and a quarter times as great as with one cell only.

Care must be taken in series connection that no cell is connected in the reverse direction, otherwise its e.m.f. becomes a back e.m.f., which is deducted and not added in the total e.m.f., while its internal resistance is still added.

EXAMPLE

Four primary cells, each of e.m.f. 1.4 V, and internal resistance 1 Ω, are connected in series to a circuit of resistance 0.4 Ω. One cell is inadvertently connected in the reverse direction (Fig. 13.5). Calculate the current in the circuit.

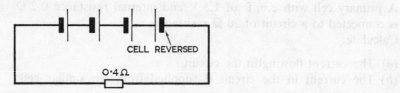

CELL REVERSED

$0.4\ \Omega$

Fig. 13.5

Total useful e.m.f. in circuit $= (3 - 1)\ 1.4 = 2.8\ \text{V}$

Internal resistance of four cells in series $= 4\ \Omega$

Total resistance of circuit $= 4 + 0.4 = 4.4\ \Omega$

Current $= \dfrac{2.8}{4.4} = 0.636\ \text{A}$

Parallel connection

When a number of similar cells are connected in parallel, all the positive terminals are connected to each other or to a common point, and the negative terminals also have a common junction, as in Fig. 13.6. The two common connections become the positive and negative terminals of the battery. We see from the figure that the e.m.f. of the battery is the e.m.f. of one cell only. The combination is equivalent to one large cell whose plates are increased in area. The internal resistance is reduced in proportion to the number of cells, and the battery under suitable conditions is able to give increased current.

Let n separate cells of e.m.f. E, and internal resistance r be connected in parallel. The e.m.f. of the battery $= E$, and the total internal resistance $= r/n$.

As with series connection, the actual increase of current depends upon the ratio of total internal resistance to the external resistance.

If the total internal resistance is small compared with the external resistance, parallel connection is of little use.

On the other hand, if the total internal resistance is large compared with the external resistance, parallel connection will give a satisfactory increase of current.

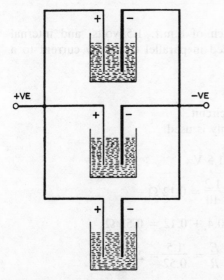

Fig. 13.6 Cells in parallel

EXAMPLE

Six Leclanché dry cells, each of e.m.f. 1.5 V, and internal resistance 0.3 Ω, are connected in parallel to supply current to a circuit of resistance 20 Ω.

Calculate:

(a) The current in the external circuit.
(b) The current if one cell only is used, the other five being disconnected.

(a)   Total e.m.f. = 1.5 V

Total internal resistance $= \dfrac{0.3}{6} = 0.05$ Ω

Total resistance $= 20 + 0.05 = 20.05$ Ω

$$I = \frac{E}{R} = \frac{1.5}{20.05} = 0.075 \text{ A}$$

(b) $I = \dfrac{E}{R} = \dfrac{1.5}{20.3} = 0.074 \text{ A}$

Note: There is practically no difference in the circuit current.

EXAMPLE

Ten Leclanché wet cells, each of e.m.f. 1.5 volts, and internal resistance 1.2 Ω, are connected in parallel to supply current to a circuit of resistance 0.4 Ω.

Calculate:

(a) The current through the circuit.
(b) The current if one cell only is used.

(a)   Total e.m.f. = 1.5 V

Total internal resistance $= \dfrac{1.2}{10} = 0.12$ Ω

Total resistance $= 0.4 + 0.12 = 0.52$ Ω

$$I = \frac{E}{R} = \frac{1.5}{0.52} = 2.88 \text{ A}$$

(b) $I = \dfrac{1.5}{0.4 + 1.2} = \dfrac{1.5}{1.6} = 0.94 \text{ A}$

Note: There is an appreciable increase of current due to the parallel connection.

Care must be taken not to connect any cell wrongly, as the result is to short-circuit the whole battery through the internal resistance of the wrongly connected cell.

Cells of different e.m.f.'s should not be connected in parallel, as the cell with the greater e.m.f. will send a local current through the cells of lower e.m.f., although the effect is less serious than in the previous case.

Series-parallel connection

Equal numbers of the cells are connected in series, and the separate sections are connected in parallel, as in Fig. 13.7. The e.m.f. of the combination is the e.m.f. of one series section. The internal resistance of the whole is the internal resistance of one series section divided by the number of sections.

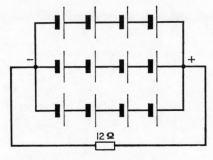

Fig. 13.7   Cells in series parallel

Let there be n cells in series per section, each of e.m.f. E, and internal resistance r. Let there be p sections in parallel.

Then, Total e.m.f. $= nE$,

and Total internal resistance $= \dfrac{nr}{p}$.

EXAMPLE

Twelve dry cells, each of e.m.f. 1.4 volts, and internal resistance 0.4 Ω, are connected in series-parallel, as in Fig. 13.7 in sections of four cells in series, and three sections in parallel. The terminals are connected to a circuit of resistance 12 Ω.

Calculate:

(a) The current flowing in the external circuit.
(b) The current flowing in each section.
(c) The potential difference of the battery.

(a) Total e.m.f. $4 \times 1.4 = 5.6$ V

Total internal resistance $= \dfrac{nr}{p} = \dfrac{4 \times 0.4}{3} = \dfrac{1.6}{3} = 0.53\ \Omega$

Current in external circuit $= \dfrac{5.6}{12 + 0.53} = \dfrac{5.6}{12.53} = 0.447$ A

(b) Current in each section $= \dfrac{\text{Total current}}{3} = \dfrac{0.447}{3} = 0.149$ A

(c) Potential difference of battery
$$= 5.6 - (0.53 \times 0.447) = 5.6 - 0.237 = 5.363 \text{ V}$$

14 Secondary cells

This type of electric cell is used for the storage of energy. When a direct current of suitable voltage is applied to the cell the electrical energy used causes chemical changes in the cell, and chemical energy is stored. This chemical action is reversible and the chemical energy stored can be released as a direct electric current.

The lead-acid cell

If two electrodes of lead are immersed in a solution of sulphuric acid, and are connected to a direct current supply of suitable voltage, a current will flow through the cell. The electrode at which the current enters the cell, the positive electrode, will gradually change colour from grey to brownish, while the negative electrode will remain unchanged in colour. The action taking place is that some of the water (H_2O) in the solution splits up, the hydrogen being liberated freely at the negative plate, while the oxygen combines with the lead of the positive plate forming a small quantity of lead peroxide on the surface.

If the terminals are connected to a resistance circuit, the cell will discharge and a current will flow in the circuit in the opposite direction to the original charging current, i.e. the current flows out of the cell by the positive terminal. Both plates will assume a greyish colour owing to the formation of a small quantity of lead sulphate on both. The cell is said to be 'formed' when sufficient quantities of the above-mentioned chemicals are formed on the plates by repeated charging and discharging.

The above actions are repeated throughout the life of the cell.

Pasted plates

The process of completely 'forming' the plates is in practice a long one. This process is shortened by the use of 'pasted' plates. The plates are made in grid form with certain oxides of lead pressed into the grids. The cell, as supplied to the customer, only requires a preliminary charge to be ready for use. Pasted plates are usually used in small cells. These plates more easily disintegrate than formed plates. For large capacity batteries it is common practice to use formed positives and pasted negatives.

Make-up of cells

The capacity of a cell depends upon the area of the plate surface, i.e. the larger the plate the bigger the capacity. To save space and for convenience, the plates are made in small sizes and a number of similar polarity are fixed together by lead-burning to a common bar which acts as a support. The positive and negative elements are interleaved, but are kept apart by separators, which may be of glass, ebonite, or of specially treated wood. The whole is set up in a suitable case of glass or ebonite. Large capacity non-portable cells are left open at the top for ease of maintenance and repair. Figure 14.1 shows such a cell with glass-rod separators. Portable cells have sealed tops through which the terminals project, and in which is fixed a filler cap, the top of which is pierced to allow gases to escape. The electrolyte is a mixture of sulphuric acid and water in such proportions that its density fully charged is about 1.21.

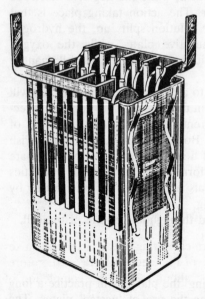

Fig. 14.1 Lead-acid secondary cell (Hart Accumulator Co Ltd)

Charging

As noted earlier, a secondary cell is charged by passing a direct current through it from positive terminal to negative terminal (see Fig. 14.2). The steady open circuit e.m.f. of a fully charged lead-acid cell is about 2.1 V. To charge the cell a voltage greater than this is needed, part to overcome the back e.m.f. of the cell, and the remainder to pass current through the cell against the internal resistance. The internal resistance of a commercial lead-acid secondary cell is normally very low. The potential difference of the cell gradually rises during charge to a maximum of about 2.7 V, when there is copious 'gassing' at the plates.

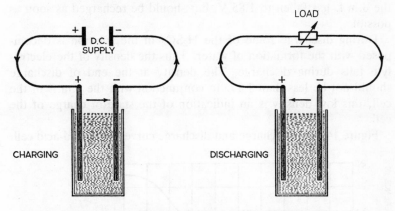

Fig. 14.2 Charging and discharging the lead-acid cell

There are two methods of charging: (1) the constant voltage method, (2) the constant current method.

In (1) the voltage is kept constant and must be above the final value of 2.7 V per cell. Obviously, at the commencement of the charge when the back e.m.f. is low, the current will be large and will gradually fall off to a minimum value at the end of charge.

In (2) the current is kept constant, either by the gradual cutting out of a resistance in the charging circuit or by increasing the charging voltage by some other form of control.

Sulphuric acid (H_2SO_4) is formed during the charge and mixes with the electrolyte. The density of the electrolyte rises, therefore, during the charge and should be about 1.21 at the completion of the charge, for large storage cells, and 1.24 for small portable cells. With a correctly mixed electrolyte the density is an indication of the state of charge of the cell. Tests of density are taken with a hydrometer during charge. The colour of the plates is also an indication of the state of charge. The positive plate should be of a chocolate-brown colour and the negative plate of a silver-grey colour.

Discharging

When the cell is being used to supply current to an external circuit, the e.m.f. falls very rapidly to 2 V. The actual potential difference depends upon the amount of current flowing. The e.m.f. then gradually falls over a period of time, which depends upon the average value of the current taken. The cell should not be used after the e.m.f. has fallen to 1.85 V, but should be recharged as soon as possible.

During discharge some of the H_2SO_4 in the solution is decomposed with the formation of water. Thus the density of the electrolyte falls during discharge. The density at the end of discharge should not be less than 1.17. In conjunction with the e.m.f. of the cell, this low density is an indication of the state of charge of the cell.

Figure 14.3 shows charge and discharge curves for a lead-acid cell.

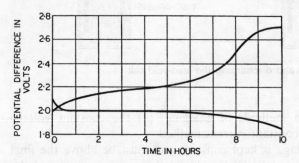

Fig. 14.3   Typical charge and discharge curves for lead-acid cell

Capacity and efficiency

The 'capacity' of a cell is given in terms of ampere-hour output. This is the product of discharge current and time and is usually specified at a definite discharge rate. Thus a cell of 120 ampere-hours at the 10-hour rate of discharge will give 12 A output continuously for 10 hours. At a lower rate of discharge something more than 120

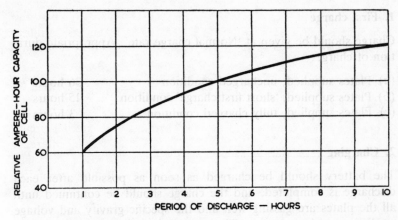

Fig. 14.4 Lead-acid cell. Curve showing how the ampere-hour output varies with the rate of discharge (Hart Accumulator Co Ltd)

ampere-hours will be available. At a greater rate of discharge the full output in ampere-hours will not be forthcoming (see Fig. 14.4).

The 'efficiency' of a cell is given in two ways: (1) ampere-hour efficiency, (2) watt-hour efficiency.

The ampere-hour efficiency is the ratio

$$\frac{\text{ampere-hours during discharge}}{\text{ampere-hours during charge}}$$

As a percentage this is about 80 per cent for a good cell.

The watt-hour efficiency is the ratio

$$\frac{\text{watt-hours during discharge}}{\text{watt-hours during charge}}$$

The watt-hour efficiency is lower than the ampere-hour efficiency, as the average potential difference during discharge is smaller than the average potential difference during charge. The watt-hour efficiency for a good cell is about 66 per cent.

It is assumed in the above that the cell after discharge is charged up to its original state.

General working instructions for storage batteries

The following instructions are extracted from working instructions by the makers of the 'Hart' lead-acid secondary cells.

1. First charge

Charge should be given at 'Normal charge rate'. Appropriate duration of charge:

(*a*) Plates supplied, uncharged condition	36 hours
(*b*) Plates supplied, 'short first charge' condition	15 hours
(*c*) Plates supplied, fully charged condition	3 hours

2. Charging

The battery should be charged as soon as possible after each discharge is completed, and the charge should be continued until all the plates are gassing well and the specific gravity and voltage readings remain constant. The normal rate of charge should be adhered to in general working, but, if necessary, the maximum rate may be given at the commencement of a charge, providing the rate is reduced to the normal as soon as the voltage of the battery reaches an average of 2.4 V per cell. The charge should then be continued until all the plates are gassing well.

The amount of excess charge that is necessary will vary according to the conditions under which the battery is worked.

An ampere-hour excess of 10 to 15 per cent over the previous discharge is usually sufficient, except when there are long intervals between the charges.

The installation of a recording voltmeter will also greatly assist the attendant in deciding when a charge is complete. During charge, the voltage of the cells rises steadily at first and more rapidly towards the end of charge until, at the end of charge, no further rise takes place.

3. Spray arresters

Spray arresters are provided with the battery and consist of small glass sheets, placed so as to cover the tops of the plates. By their use, evaporation of the electrolyte will be greatly retarded, and the loss by spray during gassing will be greatly reduced. The sheets should be cleaned occasionally to remove any dust or dirt collected on their surface. They are not required for cells with closed tops. (In connection with this paragraph the author would draw attention to the use of a layer of oil on top of the electrolyte in open cells, to reduce spraying and evaporation.)

4. Level of electrolyte

The level of electrolyte must never be allowed to fall below 12 mm above the top of the plates, as it is essential that the plates be kept well covered. Distilled water must be added to compensate for evaporation losses.

5. Testing

The condition of each cell should be tested at least once a week by means of a cell-testing voltmeter. This test should be made at the end of discharge, when the voltage per cell should not be lower than 1.85 V, the current being at the 10-hour discharge rate. The voltage of each cell should also be tested at end of charge and the readings taken about 10 minutes before current is stopped. The specific gravity of electrolyte in each cell should be tested once a week at end of charge and, at the same time, the plates in every cell should be carefully examined. If any internal contacts are found between adjacent plates, they must be at once removed.

6. Specific gravity

The specific gravity of the electrolyte when the cells are fully charged should be from 1.200 to 1.210, and when fully discharged, at the 10-hour rate, about 1.170. When specific gravities are referred to in these instructions it is intended that the readings should be regarded as being correct at normal temperature (16 °C), and if the temperature differs from this, the necessary correction should be made. An increase in temperature above 16 °C will result in the specific gravity becoming lower, the reduction being 0.0006 specific gravity for 1 °C rise. In a similar way the specific gravity will rise 0.0006 for every 1 °C fall. For example, if the specific gravity of the electrolyte is 1.205 at 16 °C, the same electrolyte will be only 1.202 at 21 °C, or 1.208 at 11 °C.

If at any time the specific gravity rises above 1.210, it must be reduced to normal value by adding distilled water.

7. Irregularity

Any cell, in which the plates do not gas freely at the end of charge, should be carefully examined for internal contacts, caused by scale, etc.; these contacts must immediately be removed and extra charge given to the cell until it is again in thorough working condition. In cases where the cells are joined together by means of removable bolts, the simplest way to give this extra charge is repeatedly to disconnect the cell out of discharge circuit, and reconnect it during charge.

8. Treatment when out of use

If the battery is to be taken out of regular use, the electrolyte must not be removed, but the battery must previously be fully charged, and care taken that the plates are well covered with electrolyte. A short charge should be given about once a fortnight till the plates gas freely, by which treatment the cells will be kept in order.

WARNING

It is dangerous to bring a flame, spark, or any ignited material near the battery at any time, as this may ignite the gases from the cells and thus cause an explosion. A spark may occur through a connector being loose, or being removed while the current is passing, and the utmost care should always be exercised in this respect. This warning applies especially to cells with closed tops.

Alkaline secondary cells

There are two main types of alkaline cell, the nickel-iron and the nickel-cadmium, the construction and action of both being very similar. The active material in the plates is contained in perforated steel compartments, which are set up in rigid steel frames. The frames or plates of the same polarity are properly spaced by steel washers and firmly bolted together, the complete assembly including a steel terminal bolt. The positive and negative plates are separated by nylon rod insulators. The whole assembly is mounted in a welded sheet-steel container holding the electrolyte, the terminals being brought out through sealing glands. The details which follow apply only to the nickel-cadmium cell.

The 'Nife' nickel-cadmium cell

Figure 14.5 shows the general construction of this cell. The active material of the positive plate consists mainly of nickel hydroxide with other ingredients, and the active material of the negative plate

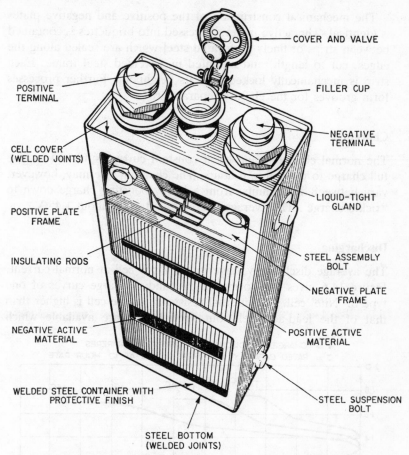

POSITIVE TERMINAL

CELL COVER (WELDED JOINTS)

POSITIVE PLATE FRAME

INSULATING RODS

NEGATIVE ACTIVE MATERIAL

WELDED STEEL CONTAINER WITH PROTECTIVE FINISH

STEEL BOTTOM (WELDED JOINTS)

COVER AND VALVE

FILLER CUP

NEGATIVE TERMINAL

LIQUID–TIGHT GLAND

STEEL ASSEMBLY BOLT

NEGATIVE PLATE FRAME

POSITIVE ACTIVE MATERIAL

STEEL SUSPENSION BOLT

Fig. 14.5 'Nife' nickel-cadmium alkaline battery (Alkaline Batteries Ltd)

is a preparation of cadmium and iron oxides. The electrolyte is a solution of potassium hydrate in distilled water with a normal specific gravity of 1.19.

The chemical action is perfectly reversible. On discharge the nickel hydroxide in the positive plate loses oxygen and is reduced to a lower form, while the cadmium and iron in the negative plate are oxidized to cadmium and iron oxides respectively. On charge the reverse action takes place, the oxides being reduced to their original metallic forms and the lower nickel hydroxide is restored to the original condition. The electrolyte does not enter into chemical combination with the plates, the specific gravity remaining constant during charge and discharge.

The mechanical construction of the positive and negative plates is identical. The active material pressed into briquettes is contained between strips of finely perforated steel which are sealed along the edges, cut to length, and mounted in a welded steel frame. Each strip is mechanically locked with its neighbour. Further processes form grooves for the ebonite separator rods.

Charging

The normal charge should be at constant current sufficient to give full charge to the cell in 6 hours. The charging rate may, however, vary between wide limits, from a heavy boosting charge down to 'trickle charge'. The average charging voltage is about 1.45 V.

Discharging

The average discharge voltage is 1.2 V per cell at normal current. Figure 14.6 gives the normal charge and discharge curves of one type of 'Nife' cell. The internal resistance of the cell is higher than that of the lead-acid cell, but special cells are available which

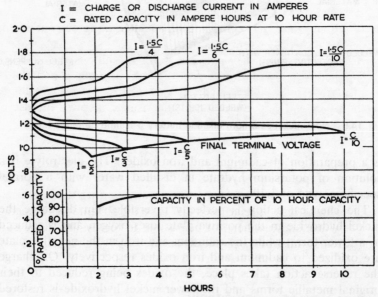

Fig. 14.6 'Nife' Cells – Type F. Charge and discharge curves showing final terminal voltage and capacity available at various rates (Alkaline Batteries Ltd)

compare very favourably and give heavy current discharge without excessive internal voltage drop. There is no limit to the amount of current which may be taken with safety and even short circuits will do no damage. No gases are produced during discharge, and the cells are sealed with vent caps rendering them unspillable. The cells are also practically free from self-discharge and may be left unattended for long periods.

Efficiency

The ampere-hour efficiency of the 'Nife' accumulator under normal conditions is from 75 to 80 per cent, with a watt-hour efficiency of 60 to 65 per cent. These values are slightly below those of the lead-acid type of battery.

Methods of charging secondary cells

Direct current is necessary, and this may be obtained in some cases straight from the supply authority's mains. Where the mains supply is alternating current, one of the following methods may be adopted:

- *Motor-generator set*, consisting of an alternating current motor driving a direct current generator.
- *Rotary convertor*, a single-armature machine, which, when supplied with alternating current through slip-rings at one end of the armature, will deliver direct current from a commutator at the other end of the armature. This usually involves the use of a special transformer.
- *Rectifiers*, of various types such as:

  (a) Metal rectifiers – copper oxide and selenium rectifiers are the two common types in use.
  (b) Semi-conductors – silicon and germanium are the two semi-conducting materials used and have the advantage, compared with metal rectifiers, of being able to rectify very large currents whilst being physically small.

Direct current mains supply

A variable resistance is necessary to control the charging current. In charging cells, either single or as a number in series, a bank of incandescent carbon lamps may be used. The lamps are connected in parallel, and switched in or out as needed.

EXAMPLE

A two-cell lead-acid battery of 20 ampere-hour capacity at the 10-hour discharge rate is to be charged from a 230-V direct current supply.

Assuming a charging rate of 1.5 A
A 16-candela, 230-V, 60-W lamp will carry 0.26 A
Therefore, six lamps in parallel will carry $6 \times 0.26 = 1.56$ A

The back e.m.f. of the cell may be neglected in this case (see Fig. 14.7).

With larger capacity cells, very many lamps would be required, and a series regulating resistance is used instead of lamps.

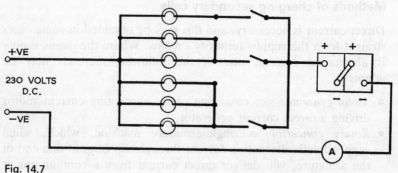

Fig. 14.7

EXAMPLE

Thirty nickel-cadmium cells, each of discharge capacity 40 A for 5 hours (or 200 ampere-hours at the 5-hour discharge rate), are to be charged at constant current for 6 hours. The supply is 230 V direct current. Calculate the value of the variable charging resistance required.

Ampere-hours required at 80 per cent efficiency

$$= \frac{40 \times 5 \times 100}{80} = 250 \text{ ampere hours}$$

Constant current for 6 hours $= \dfrac{250}{6} = 42$ A approx.

Beginning of charge:

Back e.m.f. of thirty cells in series at beginning of charge

$$= 30 \times 1.35 = 40 \text{ V}$$

(See Fig. 14.6 for voltages at beginning and end of charge.)
Potential difference available = 230 − 40 = 190 V
Series resistance required

$$R = \frac{V}{I} = \frac{190}{42} = 4.52 \ \Omega$$

End of charge:

Back e.m.f. of thirty cells = 30 × 1.7 = 51 V

Potential difference available = 230 − 51 = 179 V

Series resistance required = $\frac{179}{42} = 4.24 \ \Omega$

The resistance would be 4.52 Ω cut down in, say, four steps of 0.07 to 4.24 Ω (Fig. 14.8).
Note: If the charging were to be done by the constant voltage method, using the 4.52-Ω resistance as a fixed resistance, the charging rate at beginning of charge would be 42 A, as already found. At end of charge the charging rate would be $\frac{179}{4.52} = 39.6$ A.

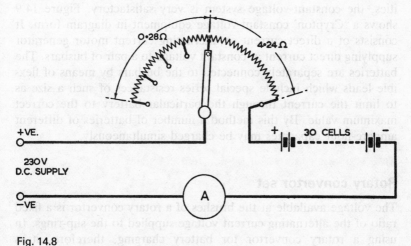

0·28Ω   4·24Ω

+VE.   + 30 CELLS −

230V
D.C. SUPPLY

−VE   A

Fig. 14.8

Motor generator set

In general, a shunt-wound generator should be used for battery charging. In the event of the supply to the motor failing during charge, say, because of blown fuses, the generator would continue to run as a shunt motor in the same direction of rotation, with no damage to the battery. A series-wound generator in the same circumstances would try to run as a series motor in a reverse direction, with the possibility of serious damage to both generator and battery. A compound-wound generator would tend to run as a motor in the same direction but at an increased speed upon a failure of the motor supply. It is common practice to connect a reverse-current circuit breaker in the battery charge circuit, which opens the circuit automatically if the battery begins to discharge into the generator.

The direct current generator is wound to give a suitable range of voltage up to the maximum that may be required. Suppose that the maximum number of 2-V lead-acid cells to be charged is fifty, then the maximum voltage of the generator would need to be (50 × 2.7) + say 10, or 145 V. A field-regulating rheostat in the generator shunt circuit would lower the generator voltage as required, either to cut down the charging rate or to enable fewer cells to be charged. Direct control of the voltage to the charging circuits by this means is more efficient than series resistance control. When a large voltage range is required, a separately excited generator is installed, since a shunt generator would be unstable at low voltages.

For the separate charging of batteries of various sizes and capacities, the constant-voltage system is very satisfactory. Figure 14.9 shows a 'Crypton' constant-voltage equipment in diagram form. It consists of a direct current or alternating current motor generator supplying direct current at constant voltage to a pair of busbars. The batteries are separately connected to the busbars by means of flexible leads which include special series resistances of such a size as to limit the current through the particular battery to the correct maximum value. By this method a number of batteries of different ampere-hour capacities may be charged simultaneously.

Rotary convertor set

The voltage available at the brushes of a rotary convertor is a fixed ratio of the alternating current voltage supplied to the slip-rings. In using a rotary convertor for battery charging, therefore, it is generally necessary to connect a tapped transformer between the

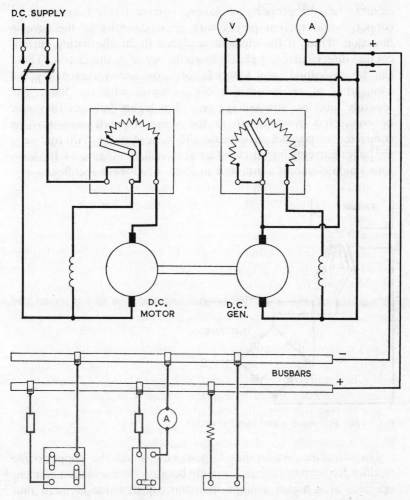

Fig. 14.9 'Crypton' constant-voltage charging equipment

supply and the convertor slip-rings to vary the alternating current voltage and thus the direct current voltage. Alternatively, a charging resistance would be used.

Metal rectifiers

The copper-oxide rectifier in its elementary form consists of a copper washer on one side of which is formed a layer of cuprous oxide. The element has the property of passing current freely from oxide to copper, while current passes with great difficulty in the reverse direction. Thus, if the element is placed in an alternating current circuit, direct current pulsations would occur in the circuit. These can be smoothed into a satisfactory continuous current by the connection of an inductance coil in series with the load. The separate units are low voltage only. For higher voltages they may be connected in series up to the value required, and may be connected in parallel to increase the current value. Alternatively, for larger currents bigger elements are available. Figure 14.10 shows four single elements connected to form a full-wave rectifier.

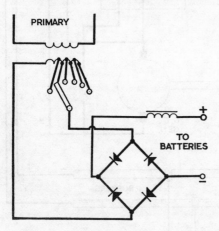

Fig. 14.10   Full-wave metal rectifier circuit

The selenium rectifier has largely superseded the copper-oxide rectifier for battery charging, mainly because each selenium element operates at a higher voltage than the copper-oxide element thus needing fewer elements for any particular voltage. The rectifier consists essentially of a steel plate coated with a thin layer of special selenium compound. A thin layer of alloy serves to make contact with the selenium compound, and rectification takes place at the intimate junction of the alloy and the compound, in so far as current flows readily from the steel plate to the alloy layer, while the junction offers a very high resistance in the opposite direction. The general make-up of elements in series and in parallel is similar to that of the copper-oxide rectifiers.

Other types of rectifiers such as the semi-conductor junction diodes have their own special qualities and uses.

Half-wave rectification

If only one rectifier unit is used in the simple circuit shown in Fig. 14.11 the current will only pass in the half cycle when the rectifier has low resistance (forward direction of current). In most practical cases if the a.c. supply voltage is not of the correct value for the circuit, the voltage will have to be adjusted using a transformer, whose secondary windings supply the rectifier and load.

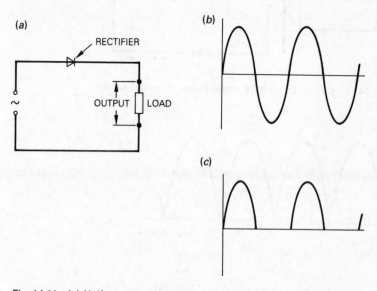

(a)

RECTIFIER

OUTPUT   LOAD

(b)

(c)

Fig. 14.11 (a) Half-wave rectification circuit (b) Input a.c. wave form (c) Output wave form. Half wave

Full-wave rectification

When the output from a rectifier unit needs to be similar to that for a steady direct current or voltage, then either two or four rectifying units have to be employed.

If two rectifiers are used then the supply is taken from a centre-tapped transformer as shown in Fig. 14.12.

One rectifier unit will rectify the positive half-cycle and the other unit will rectify the negative half-cycle. Although the output will still be a series of pulses, these pulses will be closer together then the half-wave circuit output and may be seen in Fig. 14.13.

An alternative method of obtaining full-wave rectification of an a.c. supply is to use a 'Bridge' circuit as shown in Fig. 14.14.

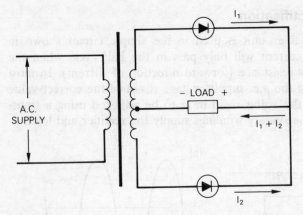

Fig. 14.12 Circuit diagram for single-phase, full-wave rectifier

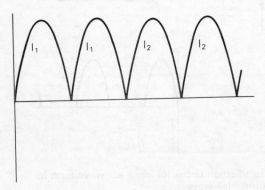

Fig. 14.13 Output waveform. Full-wave

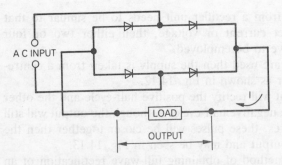

Fig. 14.14 Full-wave bridge rectifier circuit

This circuit requires four rectifier units but has the advantage of not requiring a bulky expensive centre-tapped transformer.

The wave form of the output will be the same as for the centre-tapped transformer circuit, and for some applications where a steady output similar to that of a battery or d.c. generator is required, the pulsating output of the rectifier must be smoothed.

The smoothing is obtained by use of either a capacitor or inductor, or a combination of both as shown in Figs 14.15 and 14.16.

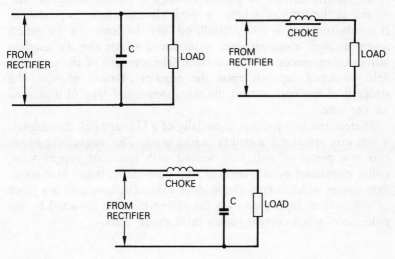

Fig. 14.15

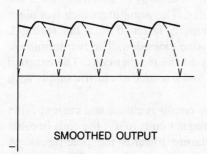

SMOOTHED OUTPUT

Fig. 14.16

15 Electric signalling systems

If an electric current is passed through a coil of insulated wire wound in the form of a helix, a weak electro-magnet is produced. If a soft-iron core is now introduced into the centre of the coil, a much stronger electromagnet is produced which may be used to attract other pieces of iron or steel. The strength of the magnetic field produced depends upon the number of turns of wire, the strength of the current and the size, shape and type of iron used for the core.

The electric bell consists essentially of a U-shaped electromagnet, a soft iron armature, a striker, and a gong. The magnet comprises two pole-pieces of soft iron wound with insulated copper wire, either enamelled or silk-covered, the pole-pieces being fixed to an iron frame, which acts as the magnet yoke. The armature is a piece of soft iron so supported by a flat spring as to be attracted by the pole-pieces when current passes through the coils.

Single-stroke electric bell

Figure 15.1 shows a single-stroke bell. AA are the two pole pieces, F is the frame or yoke of soft iron, to which the two pole-pieces are fixed. The yoke is made with extensions to which are fixed the armature spring and the gong pillar. The armature which is a piece of soft iron of rectangular section, is fastened to a flat spring L which in turn is fastened to the yoke extension. The two terminals, T_1 and T_2, are connected directly to the magnet coils. The external circuit includes a push or switch, and a source of electric supply such as a battery of primary cells.

When the push is operated, the circuit is closed and current flows in the bell from T_1 through the magnet coils to T_2; the coils become magnetized and attract the armature towards the pole-pieces. A striker attached to the armature hits the gong once. The armature

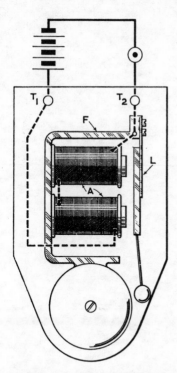

Fig. 15.1 Single-stroke electric bell

remains in the attracted position as long as the current flows, that is, as long as the bell-push is pressed. The circuit must be opened and remade before the gong can be again struck. The position of the armature is so adjusted that the striker remains just clear of the bell after the stroke to prevent muffling of the sound.

This type of bell is frequently used on railways and in mines, as it can be used for signalling purposes, It may also be used in many other situations where a longer signal would disturb clerical workers.

Trembler bell

This type of bell is essentially similar to the single-stroke bell. Figure 15.2 shows a trembler bell with its external circuit. A flat spring L is attached to the back of the armature, and in the rest

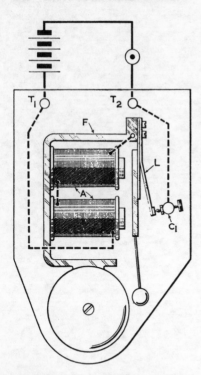

Fig. 15.2 Electric trembler bell

position bears against an adjustable contact screw C_1. The connection from T_2 is brought to C_1, and connection is made from T_1 to the magnet coils. When the bell-push is operated, current flows from T_1 through the magnet coils and along the spring L to C_1 and back to T_2.

When the current flows the armature is attracted, the gong is struck, the spring is drawn away from the contact screw C_1, and the circuit is opened. The cores become demagnetized and the armature returns to the original position. The circuit is once again made at C_1 and the bell is again struck. The cycle of operations is repeated constantly until the circuit is opened at the push. The frequency of striking the bell may be controlled to a little extent by adjusting the width of the contact gap. The contacts at C_1 are of some material as silver or platinum which does not oxidize easily.

The trembler bell is the one most commonly used as a door bell for houses and offices.

Continuous ringing bell

This is a trembler bell with mechanical and electrical arrangements such that the bell continues ringing after the bell-push has been released; Fig. 15.3 shows the details.

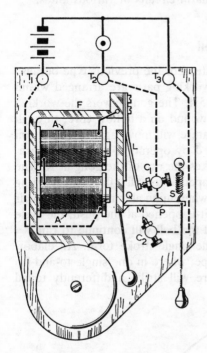

Fig. 15.3   Continuous ringing bell

There are three terminals connected to the outside circuit T_1, T_2, and T_3. The bell as connected to terminals T_1 and T_2 is the normal trembler bell. An extra terminal T_3 is provided, connected externally directly to the battery terminal, and internally to a second contact pillar C_2. A metal arm M pivoted at P and controlled by a spring S rests normally on a catch Q on the armature. A connection is made from C_1 to the arm M at P. The circuit through the bell from T_1 to T_3 is broken at C_2. When the push is operated the armature is attracted by virtue of the operation of the circuit T_1 to T_2. The catch Q releases the arm M, one end of which is drawn upwards by the spring S, whilst the other end moves downwards to make contact with C_2. The circuit T_1 to T_3 is now closed, and the bell operates continuously regardless of the opening of the circuit $T_1 T_2$ at the bell-push. To stop the bell ringing, the catch must be reset by pulling the string or chain hanging from M, thus breaking the continuous ringing circuit.

This type of bell is useful for alarm circuits of various kinds.

Circular form of trembler bell

This type of bell is often used instead of the previous type because of its more compact form. The working parts are arranged within a circular framework, as in Fig. 15.4. The gong covers the working parts. A single magnet coil A is wound on a hollow core, through which a rod-shaped soft iron armature will move against a light coil spring when the coil is magnetized by operation of the bell push. One end of the armature rod strikes the rim of the gong at K. The other end of the armature is so shaped as to pull a flat contact spring L and so open the contacts at C_1, whereupon all parts return to their original positions. Until the circuit is opened at the push, the cycle of operation will be repeated and the bell will continue to ring.

This bell can be obtained in the single-stroke form. An elaboration of this is the 'two-chime' type, where in one single to-and-fro motion each end of the armature will strike a differently toned sounding tube.

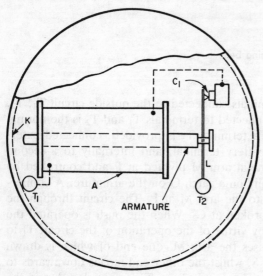

Fig. 15.4 Circular form of bell

Buzzer

A buzzer is often used in an office or in a similar situation as a quiet signal. In construction the buzzer is similar to the trembler bell with striker and gong omitted. The armature is much lighter however and its movement is very small, giving a fairly high-pitched buzz.

Bell-pushes and contacts

In the main, bells are operated by means of bell-pushes, in preference to the use of switches, which are liable to be left on. The push contains two contacts, one fixed and one movable. The movable contact is either spring-controlled or is itself in the form of a spring, and is operated by a press-knob or button.

Bell contacts are used for such purposes as ringing a bell upon the opening of a shop door. They may be fixed above the door, with the contacts so arranged that they are pressed together when the door opens, closing the circuit and ringing the bell. In other cases they are arranged below a portion of loose flooring so that the contacts are closed when a caller stands upon the floor. Details of the various types of push and contact may be obtained from the manufacturers' catalogues.

Owing to the effect of self-induction in the bell magnet coils, a spark occurs at the make-and-break contact when the contact is broken. This effect is dangerous in gaseous or flammable situations. In these special instances flame-proof bells should be installed. This requirements also applies to bell-pushes and relay contacts.

Bell transformers

The electric bell will operate not only from direct current, but also from alternating current. When a direct current passes through the coils of an electromagnet, the magnetic poles formed have fixed polarity according to the direction of the current. If the current is reversed in direction, the poles have their polarities changed, the N pole becoming the S pole and vice versa. In each case, however the action of the magnet is to attack the armature in an attempt to shorten the magnetic path of the flux. This being so, the bell will work on reversing or alternating current. To save the expense and trouble of renewing primary cells or changing accumulators, bell circuits may be connected to the electricity supply mains through a step-down transformer. With alternating current, bells with laminated cores and yokes are preferable.

Bell transformers are cheap and economical to run. The standard primary or mains voltage is from 200 to 250 V, with secondary tappings giving, for Class A type, 4, 8 and 12 V, and for Class B, 6 V.

Indicators

When a single bell is to be rung from a number of bell-pushes in separate rooms, it is necessary that the place of origin of each call should be known. A visual indicator board is therefore installed. The board consists of a number of indicator elements, each of which is connected into one of the contact circuits. The movement of the indicator element may be seen through a hole in a glass screen, each hole being marked or painted with the respective name, e.g., front door, dining-room, etc. Figure 15.5 shows a three-indicator board in diagrammatic form.

There are various types of element: the pendulum type, the shutter type, and the luminous type.

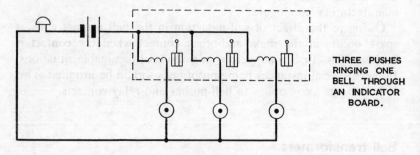

THREE PUSHES RINGING ONE BELL THROUGH AN INDICATOR BOARD.

Fig. 15.5   Three pushes ringing one bell through an indicator board

Pendulum element

Figure 15.6 is a sketch of this type of element. The element comprises a single-pole electromagnet with armature. The armature is pivoted to swing freely and a coloured flag is attached to the lower end. When the bell is rung, current flows in the circuit and through the magnet coil. The armature is drawn up to the magnet pole and is released on the circuit being opened. The armature is so weighted

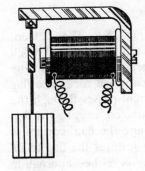

Fig. 15.6

and balanced as to swing for a considerable time. No replacement device is needed in this type of indicator.

Shutter element (mechanical replacement type)

Figure 15.7 shows this element in diagrammatic form. It consists of a single-core electromagnet with an armature in the form of a lever pivoted at A and held in normal position by a spring S. The armature is provided with a catch C which holds the flag arm in the 'off' position. The arm is pivoted at P, and is so weighted as to drop when released. When the bell is rung, current flows in the magnet, causing the armature to be attracted. The catch releases the flag arm, which falls into the alarm position. The indicator is reset mechanically by the horizontal movement of the resetting bar from right to left, the flag arm returning to its original position.

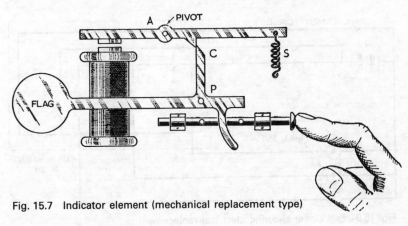

Fig. 15.7 Indicator element (mechanical replacement type)

Shutter element (electrical replacement type)

A sketch of a modern type is shown in Fig. 15.8. The element consists of two magnet coils, one of which is in series with the bell circuit. The second coil is connected in the replacement circuit. The armature is pivoted at its centre about which it will rock. The flag arm is attached to the armature. When current passes through the alarm circuit, one end of the armature is attracted, causing the flag to overbalance and fall to one side. To restore the flag, current is passed through the replacement coil, which restores the flag to its original position. This element is suitable for either alternating current or direct current working. Figure 15.9 shows a suitable circuit in diagram form.

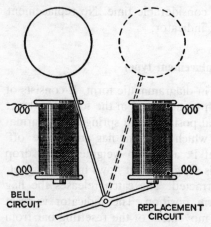

BELL CIRCUIT

REPLACEMENT CIRCUIT

Fig. 15.8 Indicator element (electrical replacement type)

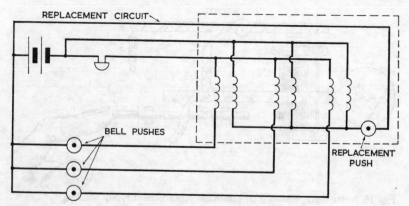

REPLACEMENT CIRCUIT

BELL PUSHES

REPLACEMENT PUSH

Fig. 15.9 Bell circuit showing electrical replacement circuit

Bell relays

When a bell circuit is a long one, the resistance of the wiring may be of such a value that the battery current may not be sufficient to ring the bell. Raising the circuit voltage by the addition of more cells may not be a satisfactory method. In such a case a relay may be fitted. The relay is similar in construction to the single-stroke bell element, having either a horse-shoe type electromagnet and spring controlled armature, or, as shown in Fig. 15.10 a single-pole electromagnet. The armature is set very closely to the poles, so that the movement is very small, requiring very little power. Figure 15.11 shows the connections of the circuit. When the distant push is operated the relay armature is drawn in, closing the local circuit which consists of bell and local battery. The bell will ring, its loudness being dependent upon the local battery power. This type of relay is also used in connection with automatic burglar alarms, fire alarms, etc.

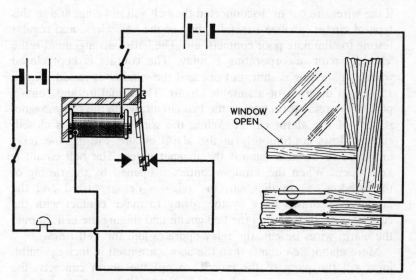

WINDOW
OPEN

Fig. 15.10   Closed-circuit burglar alarm

Burglar alarms

The simple type is the normally open circuit arrangement whereby the alarm bell is rung when a door-type contact is closed by the opening of a window or door. The objections to this circuit are that

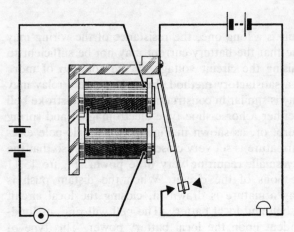

Fig. 15.11   Circuit, including bell relay

if the wires are cut or disconnected the bell will not ring, and so this type of circuit involves careful hiding of the bell wires, and regular testing to eliminate poor contacts, etc. The better arrangement is the 'closed circuit' incorporating a relay. The contact is kept closed when the window is shut, but opens if the window is raised. Figure 15.11 is a diagram of a suitable circuit. The armature and contact post of the relay form part of the bell circuit, while the relay magnet is part of the alarm circuit. While the window contact is closed, current flows continuously in the alarm circuit, causing the relay armature to be held against the magnet poles. The bell circuit is kept open. When the window contact is opened by the raising of the window, as in diagram, the relay is demagnetized and the armature is drawn back by the spring to make contact with the contact post, thus closing the bell circuit and ringing the bell. Should the alarm wires be cut, the relay operates and the bell rings.

More elaborate circuits than the above are used in lock-up buildings. All the parts of the circuit, except the alarm contacts, the wiring and the outside loud-ringing alarm bell, are contained in a control cabinet. A red signal lamp indicates when the circuit is switched off during daylight hours, and a 'delay-setting switch' may be provided to prevent the alarm sounding while the occupier is leaving the premises after setting the alarm. Other circuit arrangements can be made to fit 'raid-alarm' pushes for daytime use in the event of an attempted hold-up.

Mains voltage bells

Loud-ringing alarm bells of various kinds may be operated directly from the supply mains, alternating or direct current. They are robust bells mounted in iron cases with sturdy terminals enclosed within the cover. Bells for alternating current have laminated magnets. Conduit entry for the wiring is provided, together with an earthing terminal.

The wiring to the bell and the bell itself must be installed in conformity with the general regulations applying to lighting and power circuits dealt with earlier.

Luminous signalling systems

Silent signalling systems are particularly useful in hospitals and for night calls in hotels and similar places. Lamp signals are used instead of bells. One system of this type is here described.

Outside each room a luminous unit is fixed, which includes a glow-lamp, a relay, and a resetting push. Upon the operation of a push inside the room, the relay operates, closing a local circuit and lighting the lamp, which remains lit until the porter or servant who

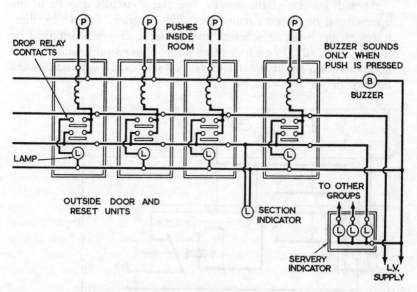

Fig. 15.12   Luminous signalling system (Gent & Co Ltd)

answers the call resets the indicator. The resetting is done by means of a mechanical push-button. The relay also closes a circuit, which lights a lamp in the porter's room or servery, indicating the group of rooms from which the call originates. A buzzer in the porter's room may be included in the installation. Figure 15.12 is a diagram illustrating the above system. The system may be extended and elaborated to any desired extent.

In this particular system the low voltage supply is 12 V direct current from accumulators, or 12 V alternating current from a suitable step-down transformer. The lamps are 3 V each. Low-voltage wiring only, as for bell systems, is necessary.

Fire alarms

All premises which come under the Factories Act must have an installation to give warning of fire.

A works system may include a number of hand-operated alarm switches, commonly of the break-glass type, automatic fire detectors at suitable points, alarm bells, and a fire indicator board.

The electrical supply for the alarm system may be mains voltage, or extra-low-voltage alternating current.

As with burglar alarm circuits, fire alarm circuits may be of the open-circuit or closed-circuit types. The diagram, Fig. 15.13, illustrates an open-circuit scheme built up from diagrams supplied by Gent & Co Ltd. On each of three floors is shown one only manual alarm and two automatic detectors. A 3-way board will indicate the floor of origin.

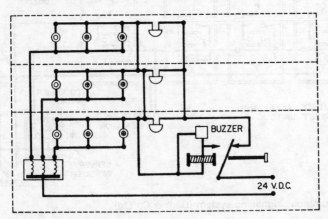

Fig. 15.13   Open-circuit fire alarms

It is an advantage to be able to silence the alarm bell after the alarm has been given, partly to relieve the batteries of a continuous load, and partly to produce a calm during which the fire may be dealt with. For this purpose, a supervisory buzzer may be included in the circuit. A push-switch arrangement causes the supervisory buzzer to sound, at the same time cutting out the loud alarm bells. When the original alarm contact is restored to its normal open-circuit position, the buzzer is automatically disconnected and the whole fire alarm system returns to its normal ready state.

Automatic fire detectors

An automatic detector will operate at any time, day or night. Briefly, a detector comprises a bi-metal strip which bends when heated to a predetermined temperature of say 66 °C. A latch is released which allows a mercury switch to tilt, and thus close the alarm circuit. The detectors are normally fixed to ceilings away from accidental or other damage, at about 6-m intervals. They must be carefully sited away from direct sunlight, hot pipes or heating appliances.

Segregation of fire alarm circuits

It was noted earlier, that all extra-low voltage circuits should be kept apart from other circuits supplied directly from the mains. In addition to this, a fire alarm installation should be completely separated from the wiring of any other circuit, and as far as possible should not follow a common route with other cables.

16 Tariffs and power-factor improvement

Although there may be some slight differences between area boards, the typical domestic tariff is comprised of a standing charge of £8.20 per quarter (13 weeks) plus 5.89p per unit (kWh). The fixed charge is based upon the floor area of the dwelling.

Commercial premises, farms, and churches

These are offered tariffs particularly applicable to their conditions. Details are obtainable from the local Area Board or supply authority.

Reduced rates are available to domestic and other consumers for the exclusive purpose of thermal storage. A time switch controls supply so that 'off peak' energy is available only between the hours of 22.00 hrs and 08.00 hrs. A typical tariff is (a) Standing charge of £2.00 per quarter plus 2.64p per unit.

Economy Seven rate

Within domestic premises a special rate is available for energy used for any seven hours between midnight and 08.00 hrs.

Tariffs for large power consumers

The two-part tariff is invariably imposed on this type of consumer. The standing charge depends upon the maximum demand of the consumer in kW or kVA, while the running charge is of the normal type. Some of the tariffs offered by the supply authorities are quite complex in their setting-out and may include a 'variation in cost of fuel' clause.

The following is a simplified example of this kind of tariff, omitting the fuel clause: £1.65 for each kVA of maximum demand in a winter month, and £0.45 for each kVA of maximum demand in a summer month, plus 0.3p per unit. (Winter months are November to February, and summer months are March to October.)

Maximum demand

Although the maximum demand is the actual maximum kW taken at any instant, it is taken in practice to mean the average demand over a fixed short period, say half an hour. It is measured by means of a Maximum Demand Indicator, which registers the total kWh taken over the prescribed period, say one month, divided by the period in hours. At the end of the given period the driving train of the indicator is automatically set to zero, leaving the slipping pointer in its highest position. If, however, during a later period the maximum demand is greater than before, the pointer is moved still further, thus the reading of the pointer is the maximum value reached during the whole time under consideration. The meter is reset manually when the meter reader makes his regular call.

EXAMPLE

A power consumer who has a fairly constant maximum demand throughout the year is offered the following tariff: £10.25 per kW of maximum demand per annum, plus 0.375p per unit. Given that his annual maximum demand is 250 kW, and his annual consumptionis 350 000 units (kWh), calculate the annual cost, and the average price per unit.

$$\text{Annual cost} = £(10.25 \times 250) + (350\,000 \times 0.375p)$$
$$= £2562.5 + £1312.5$$
$$= £3875$$

$$\text{Average price per unit} = \frac{£3875}{350\,000}$$
$$= \frac{387\,500}{350\,000}$$
$$= 1.11p$$

Further consideration of this example by the reader will indicate the financial advantages to be obtained by arranging that short-time heavy loads in various parts of a works be staggered as much as possible so as to reduce the maximum demand. The annual maximum demand given is almost twice the value of the annual average demand of a year comprising 48 hours weekly for 50 weeks. If the maximum could be reduced by load staggering to be nearly equal to the average demand, the saving would be of the order of £1000 per annum.

Power factor

The majority of industrial loads have power factors of less than unity. In many cases where induction motors predominate, the power factor is of the order of 0.7 lagging. Many authorities penalize a consumer for low power factor by means of a power factor clause in the tariff agreement. The following is an example of such a clause: the amount payable for each kW of maximum demand shall be increased by 1 per cent for each 0.01 by which the average lagging power factor for the year of supply is less than 0.9.

EXAMPLE
Assume that the consumer in the last example has an annual power factor of 0.7 lagging and is charged £10.25 per kVA of maximum demand, plus 0.375p per unit. Calculate the annual cost, and the average price per unit.

$$\text{kVA of max. demand} = \frac{\text{kW}}{\text{power factor}} = \frac{250}{0.7} = 357.14 \text{ kVA}$$

$$\begin{aligned}
\text{Annual cost} &= \pounds(10.25 \times 357.14) + (350\,000 \times 0.375\text{p}) \\
&= \pounds3661 + \pounds1312.5 \\
&= \pounds4973.5
\end{aligned}$$

$$\begin{aligned}
\text{Average price per unit} &= \frac{\pounds4973.5}{350\,000} \\
&= \frac{497\,350}{350\,000} \\
&= 1.42\text{p}
\end{aligned}$$

Compare this with the previous example, where the maximum demand charges only relate to kW.

Power-factor improvement

Electrical energy supplied at a poor power factor is more costly to the supply authority than the same energy supplied at or near unity power factor, because it entails larger currents and therefore bigger alternators, transformers, switchgear, cables, etc. In effect, all the electrical apparatus pertaining to generation and supply are bigger than need be, and bigger copper losses are sustained throughout the system owing to the increased currents. The mechanical side of generation is not affected. It is for these reasons that the supply authority includes the power factor clause in the tariff.

The consumer obviously, then, has a financial interest in improving his power factor. In the parallel last two examples, it is shown that the annual extra cost to this consumer is considerable, and he can therefore afford to spend money on apparatus which will improve the power factor.

Figure 16.1 is a simple phasor diagram of a current I lagging by angle $\phi°$ behind the voltage V. The power factor is the cosine of the angle between the current and the voltage, or $\cos \phi$. The current may be considered as comprising two components, the power component $I \cos \phi$ in phase with the voltage, and the reactive component $I \sin \phi$, lagging by 90°. If a leading current equal and opposite to $I \sin \phi$ is injected into the circuit, the two vertical components will cancel each other, and the result will be as shown in Fig. 16.2 (a) where the resultant current is in phase with the voltage, that is, at unity power factor.

It should be noted here that it is not always considered necessary to convert a lagging power factor to unity. This is because the costs of power-factor improvement beyond say 0.9 lagging may outweigh the tariff savings that could be made. Figure 16.2 (b) shows the effect on the resultant current when a leading reactive current of less than the full value of the lagging component is injected into the circuit. In this case the overall power factor is the cosine of the angle ϕ_2.

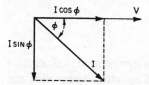

Fig. 16.1 Phasor diagram representing power factor

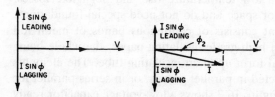

Fig. 16.2 (a) and (b) Phasor diagrams representing power-factor correction

There are various ways of supplying the necessary leading current using:

(a) Phase advancers
(b) Synchronous motors, running over-excited
(c) Capacitors.

Phase advancers

These are special machines which are connected in parallel with individual motors or other inductive apparatus, and are switched on or off with the motor. The advantage of connection at this point is that the power factor is improved in the consumer's cables and switchgear up to the motor terminals.

Synchronous motors

The synchronous motor with its direct current field over-excited, will take a current which leads the voltage by an angle which depends upon the design of the machine, and upon the amount of over-excitation. The synchronous motor when used for this purpose can either be run idle, or can be used to supply mechanical power at constant speed. The usual practice is to have a large motor connected in parallel with the works busbars. For details of the theory of synchronous motors and also of phase advancers, refer to textbooks on electrotechnology.

Capacitors

The use of capacitors for power-factor improvement has certain advantages over the foregoing methods. Capacitor assemblies are made for all voltages up to 33 000 volts, and can therefore be connected to circuits of any voltage without the use of special transformers. They have a low temperature rise and negligible losses. They occupy little floor space and do not need special foundations.

A capacitor element consists of continuous bands of metal foil separated by layers of high-grade insulating paper, the whole being wound into cylindrical form upon an insulating tube. The elements are themselves connected in parallel groups, or in series-parallel for the higher voltages. Figure 16.3 shows a low-voltage capacitor ready for immersion in oil in its transformer type tank.

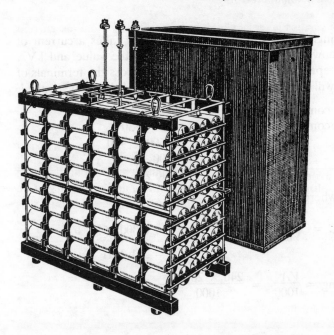

Fig. 16.3   Industrial type capacitor (Johnson & Philips Ltd)

Installation of capacitors.   For individual control, capacitors should be placed as near the load as possible (see Fig. 16.4). They should be connected across the terminals of the induction motor or other induction apparatus, under the control of the motor switch. When the motor is switched off, the capacitor will discharge itself through the motor windings. For group control, the capacitors are connected to the group busbars, and are controlled by separate switches, fitted with high ohmic resistances for discharge purposes.

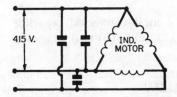

Fig. 16.4   Capacitors connected to 3-phase motor

EXAMPLE

A 240-V, single-phase, 50 Hz, induction motor takes a current of 40 A at a power factor of 0.75 lagging. Find the value and kVA rating of a capacitor which, when connected across the terminals of the motor, will raise to unity the power factor of the total supplied.

Energy component of current $= 40 \times 0.75 = 30$ A
Reactive component of current $= \sqrt{40^2 - 30^2}$
$= \sqrt{1600 - 900}$
$= \sqrt{700}$
$= 26.46$ A

$I = VC\omega$, where $\omega = 2\pi f$

therefore $C = \dfrac{I}{V\omega} = \dfrac{26.46}{240 \times (\pi \times 50)} = 0.000\ 351\ 2$ F
$= 351.2\ \mu$F

kVA rating $= \dfrac{VA}{1000} = \dfrac{240 \times 26.46}{1000} = 6.35$ kVA

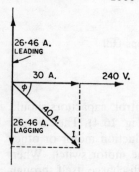

Fig. 16.5 Phasor diagram

EXAMPLE

A 415 V, 3-phase, 50-Hz, induction motor having an output of 74.6 kW runs on full load at a power factor of 0.7 lagging, and with an efficiency of 85 per cent.

Find the capacitance per phase of a mesh-connected capacitor necessary to raise the power factor: (a) to unity, (b) to 0.9 lagging.

Work to phase values

Power output of motor per phase $= \dfrac{74\ 600}{3}$
$= 24\ 867$ W

$$\text{Watts input per phase} = \frac{24\ 867 \times 100}{85}$$
$$= 29\ 255\ \text{W}$$

(a) *At unity power factor:*

$$\text{Input VA per phase} = \frac{29\ 255}{0.7} = 41\ 793\ \text{VA}$$

$$\text{Actual current per phase} = \frac{41\ 793}{415} = 100.7\ \text{A}$$

$$\text{Energy component of current} = 100.7 \times 0.7 = 70.5\ \text{A}$$

$$\begin{aligned}\text{Reactive component of current} &= \sqrt{100.7^2 - 70.5^2}\\ &= \sqrt{10\ 140 - 4970}\\ &= \sqrt{5170} = 71.9\ \text{A}\end{aligned}$$

Therefore, leading current required $= 71.9$ A

But, $I = VC\omega$, therefore $C = \dfrac{I}{V\omega}$

Therefore, $C = \dfrac{71.9}{415 \times (2\pi \times 50)} \times 10^6 = 551.4\ \mu\text{F}$ per phase

(b) *At 0.9 power factor lagging:*

Actual current per phase at 0.9 lag. $= \dfrac{70.5}{0.9} = 78.3$ A

$$\begin{aligned}\text{Reactive component} &= \sqrt{78.3^2 - 70.5^2}\\ &= \sqrt{6132 - 4970}\\ &= \sqrt{1162} = 34.1\ \text{A}\end{aligned}$$

Therefore, leading current required $= 71.9 - 34.1 = 37.9$ A

(see Fig. 16.6)

Therefore, $C = \dfrac{37.8}{415 \times (2\pi \times 50)} \times 10^6 = 290\ \mu\text{F}$ per phase

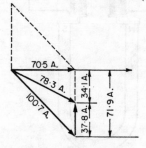

Fig. 16.6 Phasor diagram

17 Telephones

The telephone is an instrument used for the purpose of converting sound waves into electrical impulses which can be transmitted over long distances and reconverted into sound. Sound is produced by waves of pressure sent out by the sounding body, and it is heard when these waves reach the ear. When sound waves impinge upon a diaphragm, it vibrates. The frequency, amplitude, and 'shape' of the vibrations depend upon the pitch, loudness, and 'quality' respectively of the sound waves.

The simple telephone is shown diagrammatically in Figure 17.1. It consists of a permanent magnet of horseshoe shape, with coils of insulated wire wound upon the poles, and a flexible iron diaphragm fixed in position near to the poles. When the diaphragm is still, the magnetic field of the magnet is unaffected. If the diaphragm moves towards the poles, as shown (exaggerated) by dotted lines, the magnetic field is both strengthened and altered in position. If the diaphragm moves away, the field is again altered in position and weakened in strength. The lines of force thus set up and collapsed will cut the magnet coils, setting up small varying e.m.f.'s which in

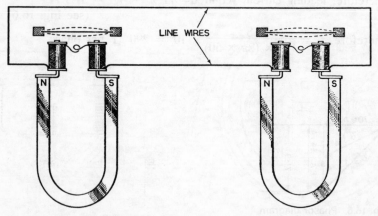

Fig. 17.1   Simple telephone circuit

turn cause varying currents to flow in the circuit. If a similar telephone is connected in series, the varying currents in the magnet coils will cause changes in the strength of the magnetic field of this instrument. The changes of field strength attract and release the diaphragm, setting up mechanical vibrations, which are converted back into sound waves of the same type as the originals. Thus, words spoken into the original telephone are reproduced in the second telephone at a distance.

The energy transmitted is small, and this arrangement is not used in practice, although it can be used in an emergency over very short distances.

The telephone receiver

In practice, the telephone instrument, already described in one form, is used as a receiver only. The modern type of receiver is shown in Figs 17.2 and 17.3.

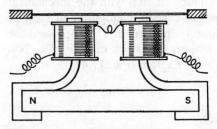

Fig. 17.2 Telephone receiver

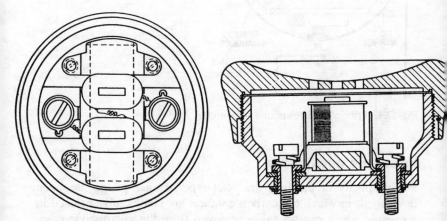

Fig. 17.3 Telephone receiver (Ericsson Telephones Ltd)

The permanent magnet is a short bar magnet with soft iron pole-pieces clamped or pressed on the ends (Fig. 17.1). These pole-pieces each carry a coil of fine insulated wire, the two being connected in series. The diaphragm is held in position very close to the pole-pieces, but does not touch them when in vibration. Figure 17.3 is a standard receiver, as made by Ericsson Telephones Ltd.

The telephone transmitter

The carbon microphone is universally used as a transmitter. It operates by virtue of the changes in resistance of carbon contacts under varying mechanical pressures. If the mechanical pressure between two pieces of carbon in contact is increased, the electrical resistance decreases. Inversely, if the mechanical pressure is decreased, the electrical resistance is increased. If, then, these two pieces of carbon in contact are connected in series with a battery, the current in the circuit will change in step with the mechanical pressure. There are several types of carbon microphone in most of which are carbon granules loosely packed and compressed or loosened by the vibration of a diaphragm, which may be of carbon or metal. The principle is shown by diagram (Fig. 17.4).

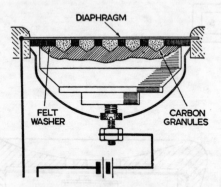

Fig. 17.4   Principles of telephone transmitter

Inset type microphone

For simplicity of upkeep, the inset type of microphone has been developed, in which the carbon contacts are completely enclosed in a capsule which can be easily removed from the microphone case.

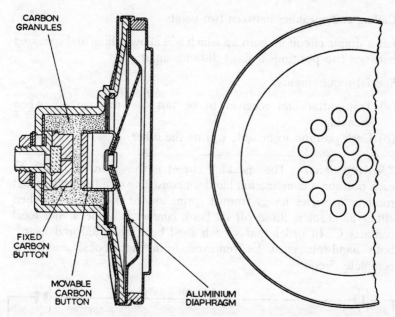

Fig. 17.5 Microphone capsule in section (Ericsson Telephones Ltd)

The complete capsule, which is enclosed in metal, is shown in section in Fig. 17.5. An insulated cylindrical pocket contains two carbon buttons – one a fixed button, whose screw protrudes through the rear of the capsule to form one terminal. The second button is fixed to an aluminium diaphragm, which vibrates under the action of speech or sound. The space between and around the button is filled with carbon granules made from anthracite coal. The circular rim of the diaphragm is fixed to the metal case which forms the second terminal. As explained above, movement of the diaphragm button varies the pressure on the granules, thus varying the resistance of the microphone.

If the microphone is connected in series with a battery and a receiver, the circuit current will vary. These current variations will be converted into sound at the receiver.

Telephone circuits

Typical circuits are described and illustrated below. There are many variations, for which specialist books should be consulted.

Calling and speaking between two points

Let a simple circuit be built up which will allow calling and speaking between two positions a short distance apart.

Special requirements:

(a) Transmitter and receiver to be out of circuit except when speaking.
(b) Each position to be able to ring the other.

Speaking circuit. The speaking circuit is shown in Fig. 17.6. At each position a combination hand-set consisting of a transmitter and receiver in series hangs from a spring-loaded metal hook. When either hand-set is lifted off its hook contact is made at the local contacts C. In order that speech shall be transmitted and heard, both hand-sets must be removed from the hooks, making a complete circuit.

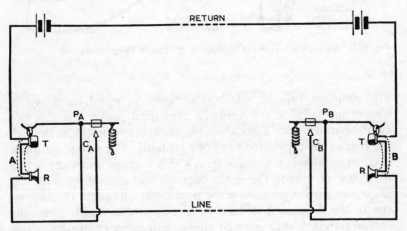

Fig. 17.6 Telephone speaking circuit

Ringing circuit. The ringing circuit is shown in Fig. 17.7 using the same pair of line wires, marked 'line' and 'return', the same receiver hooks, and the same local batteries. The additional equipment at each station comprises an extra contact C_1 on the receiver hook, only broken when the receiver is lifted, a ringing key with double contact, and a bell. The diagram showing station A ringing station B is easily followed. The circuit is: ringing key K at A, contact C_1A, along receiver hook to PA, along 'line' to receiver hook at PB,

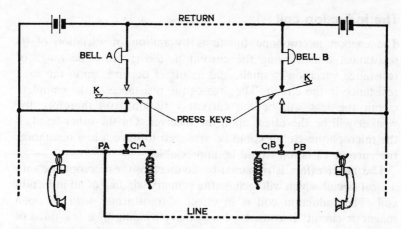

Fig. 17.7 Telephone ringing circuit

through contact C_1B to ringing key at B, to bell B, along 'return' to battery A, and to ringing key at A.

Complete circuit. Figure 17.8 shows the above two circuits combined, with station A in conversation with station B. Assuming that the separate circuits have been properly followed and understood, no further description is necessary.

The above type of circuit is suitable for short distances only, and is the type used for domestic telephones. For long distances induction coils should be incorporated in the circuit.

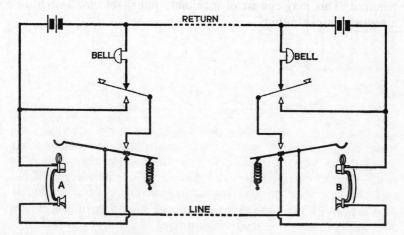

Fig. 17.8 Complete telephone circuit

The induction coil

The carbon microphone functions by reason of variation of its resistance, thus varying the current in the circuit. The range of resistance variation is small, and its effect depends upon the total resistance in the circuit. The greater the resistance of the complete circuit the less will be the current variation, and therefore the weaker will be the effect upon the receiver. On the other hand, if the microphone circuit could be arranged to have a low resistance, the current variation would be appreciable.

The latter effect is obtained by connecting the microphone into a local circuit which will include the primary winding of an induction coil. The induction coil is in effect a transformer with an open magnetic circuit. It consists of a primary winding of a few turns of insulated wire wound on an iron core, and a secondary winding of many turns of insulated wire wound outside the primary winding. The core consists of a bunch of iron wires. The induction coil eliminates from the line the direct current component of the current in the microphone.

Intercommunication telephone systems

For telephonic communication between any two of a number of stations, for instance, between the offices and shops of a works, intercommunication systems are employed. Each telephone station must be provided with a means of selecting the particular station required. This may consist of a suitable radial selector switch, or a series of press-buttons.

Index